Navigating the Math Major
Charting Your Course

AMS / MAA | CLASSROOM RESOURCE MATERIALS

VOL 73

Navigating the Math Major

Charting Your Course

Carrie Diaz Eaton
Allison Henrich
Steven Klee
Jennifer Townsend

Providence, Rhode Island

Classroom Resource Materials Editorial Board

Cynthia J. Huffman, Editor

Judith Covington   Maria Fung
Russell Goodman   Joel K. Haack
Brian Hollenbeck   Gizem Karaali
Jessica M. Libertini   Sarah Loeb
Candice Price   Matthew L. Wright

2020 *Mathematics Subject Classification*. Primary 00-01, 00A05, 01A80.

Compass cover photo by Creativeye99 / E+ / Getty Images.
Map cover photo by H20addict / E+ / Getty Images.

For additional information and updates on this book, visit
www.ams.org/bookpages/clrm-73

Library of Congress Cataloging-in-Publication Data
Names: Eaton, Carrie Diaz, author. | Henrich, Allison K., 1980- author. | Klee, Steven, author. | Townsend, Jennifer, 1988- author.
Title: Navigating the math major : charting your course / Carrie Diaz Eaton, Allison Henrich, Steven Klee, Jennifer Townsend, authors.
Description: Providence, Rhode Island : MAA Press, an imprint of the American Mathematical Society, [2024] | Series: Classroom resource materials, 1557-5918 ; volume 73 | Includes bibliographical references.
Identifiers: LCCN 2024009447 | ISBN 9781470475833 (paperback) | ISBN 9781470477752 (ebook)
Subjects: LCSH: Mathematics–Vocational guidance. | AMS: General – Instructional exposition (textbooks, tutorial papers, etc.). | General – General and miscellaneous specific topics – General mathematics. | History and biography – History of mathematics and mathematicians – Sociology (and profession) of mathematics.
Classification: LCC QA10.5 .E28 2024 | DDC 510.71/1–dc23/eng/20240324
LC record available at https://lccn.loc.gov/2024009447

Copying and reprinting. Individual readers of this publication, and nonprofit libraries acting for them, are permitted to make fair use of the material, such as to copy select pages for use in teaching or research. Permission is granted to quote brief passages from this publication in reviews, provided the customary acknowledgment of the source is given.

Republication, systematic copying, or multiple reproduction of any material in this publication is permitted only under license from the American Mathematical Society. Requests for permission to reuse portions of AMS publication content are handled by the Copyright Clearance Center. For more information, please visit **www.ams.org/publications/pubpermissions**.

Send requests for translation rights and licensed reprints to **reprint-permission@ams.org**.

© 2024 by the authors. All rights reserved.
Printed in the United States of America.

∞ The paper used in this book is acid-free and falls within the guidelines
established to ensure permanence and durability.
Visit the AMS home page at **https://www.ams.org/**

10 9 8 7 6 5 4 3 2 1   29 28 27 26 25 24

We dedicate this book to our students.

Contents

Acknowledgments   xi

How to Use This Book   xiii
    0.1   For Students   xiii
    0.2   For Instructors   xv

About the Authors   xvii

Part 1   Exploring Your Interests   1

    1   Introduction: Start Here   3
       1.1   Identity   4
       1.2   Values: What's Important to You?   6
       1.3   Constraints and Obligations   7
       1.4   Goals and Interests   8
       1.5   Roadmap for the Rest of the Book   10

    2   Planning Your Course of Study   13
       2.1   Introductory Mathematics Core   13
       2.2   Upper-Level Mathematics Courses   16
       2.3   Programs of Study   21
       2.4   Changing Your Path   26
       2.5   Reflection   27
       2.6   Lee Johnson: From Green River Community College to NASA   27

    3   Extracurricular Explorations   31
       3.1   Research Experiences   32
       3.2   Summer Internships   43
       3.3   Study Abroad Programs   44
       3.4   Math Contests   45
       3.5   Department-Level Engagement   47
       3.6   Community Engagement   49
       3.7   Scavenger Hunt   51

Part 2   Supporting Your Success   55

    4   Failure and Growth   57
       4.1   Overcoming Failure   57
       4.2   Mythical Genius (and Why You Shouldn't Care About It)   61
       4.3   Finding Community   62

4.4	Getting Back to Work	64

5 Networks and Communities of Support — 65
- 5.1 Mathematical Communities: People Create Mathematics — 66
- 5.2 Professional Societies — 67
- 5.3 Conferences and Events for Communities of Mathematicians — 72
- 5.4 Support from Your School — 73
- 5.5 Tips on Conferences and Networking — 73
- 5.6 Mathcrostic — 77

6 Technical Skills — 81
- 6.1 Collaboration Skills — 81
- 6.2 How to Read a Math Textbook — 83
- 6.3 How to Read a Math Research Paper — 84
- 6.4 Writing Math: LaTeX — 86
- 6.5 Technical Writing Skills — 87
- 6.6 How to Give a Math Talk — 91
- 6.7 How to Give a Poster Presentation — 95
- 6.8 Programming Skills — 98
- 6.9 Conclusion — 100

Part 3 Life After Graduation — 103

7 Careers for Math Majors — 105
- 7.1 What Can You Do with a Math Major? — 106
- 7.2 Data-Oriented Careers — 107
- 7.3 Research in Industry — 113
- 7.4 Careers in Finance — 116
- 7.5 Actuarial Science — 118
- 7.6 Careers in Software Engineering — 120
- 7.7 Government Careers — 122
- 7.8 Jobs in Education — 128
- 7.9 Other Career Options — 130

8 Applying for Jobs — 145
- 8.1 What Does Work-Life Balance Look Like for You? — 145
- 8.2 Preparing for the Job Hunt: Finding a Mentor — 146
- 8.3 When to Apply to Full-Time Industry Positions — 147
- 8.4 What Skills Are Useful for Non-Academic Careers? — 147
- 8.5 What If I'm Not Qualified for the Job? — 148
- 8.6 How to Write a Personal Statement or Cover Letter — 148
- 8.7 How to Write a CV or Résumé — 148
- 8.8 Online Presence — 150
- 8.9 What If I Apply and Don't Hear Back? — 150
- 8.10 I Landed an Interview! How Do I Prepare? — 152
- 8.11 Flawed AI Résumé Review — 154
- 8.12 A Social Media Silly Story — 155

9 Graduate School in the Mathematical Sciences — 157
- 9.1 Graduate Programs: Frequently Asked Questions — 157
- 9.2 Deciding Where to Apply — 159

9.3	Applying for Graduate School: Your Personal Statement	161
9.4	Letters of Recommendation	162
9.5	Standardized Tests for Graduate Admissions	162
9.6	Resources for Graduate Students	164
9.7	Amzi Jeffs and the NSF Graduate Research Fellowship	165

Conclusion 169

Bibliography 171

Acknowledgments

We would like to thank Omayra Ortega and Anisah Nu'Man for contributing their ideas to the initial development of this book. We also want to express our deep gratitude to Cynthia Huffman and the other members of the Classroom Resource Materials editorial board for the significant work they did to provide us feedback that helped us improve this book. Others we'd like to thank are Brian Fischer as well as the student draft reviewers at Bates College: Quinn Macauley, Jianing Ni, and Annie Doig. Thank you also to everyone who shared their wisdom by contributing interviews throughout the book. Finally, this book would not exist without Steve Kennedy's encouragement. We are tremendously indebted to him.

How to Use This Book

0.1 For Students

Congratulations for picking up *Navigating the Math Major: Charting Your Course*! We hope you won't be able to put it down! While you might take all the same courses as another math student, we know that there is no "one size fits all" math major. Not only the courses you take, but also the extracurricular activities you participate in, the communities you become a part of, and ultimately what you do after completing your math degree all depend on one critical factor: YOU! Your goals, your interests, and your passions are what drives your math major. The goal of this book is to give you more information about the options that are available to you so you can prepare yourself to achieve your goals.

Because of this, there might be different sections of the book that will be more useful to you at different parts of your academic career. We will divide our advice very broadly into two pieces: advice for students who are new to the math major (first/second year students or students who just declared their math major) and advice for students who are already math majors and who are thinking about what to do in their last 1–2 years of college and post-graduation.

0.1.1 For students who are new to the math major.
Maybe you just declared your math major or you're thinking about becoming a math major and you're in an exploratory phase where you want to know what your options are. In that case, the following chapters might be useful:

- Chapter 1, "Introduction: Start Here," includes some prompts to help you think about your goals and passions.

- Chapter 2, "Planning Your Course of Study," gives a high-level overview of different degree options and math classes you might take.

- Chapter 3, "Extracurricular Explorations," has information about extracurricular activities. This is a good chapter to at least skim for now, but you may skip over the sections with concrete advice about applying for research programs/internships or what to do once you're accepted. It's never too early to start thinking about summer research/internship opportunities, though, even if you won't be applying for another year or two.

- The scavenger hunt in Section 3.7 is a good way to acquaint yourself with the resources that exist on your campus to help support you in your major.

- Chapter 4, "Failure and Growth," is worth reading and revisiting as needed. Math can be hard. The struggle is sometimes very real. This chapter includes resources to help you find support and community when you need it.

- The beginning of Chapter 5, "Networks and Communities of Support," has information about professional societies in the mathematical sciences, along with other groups that you might consider joining, especially if you are from a group that has been historically marginalized in STEM. This chapter also includes information on finding and developing community with your fellow mathematicians and how to attend conferences (and *which* to attend), which is an experience we recommend even if you are a first- or second-year student.

- Chapter 6, "Technical Skills," may be beyond what you need to know right now, but you may find value in reading the first section on developing good collaboration skills. We also recommend skimming the section headings for the rest of the chapter to familiarize yourself with the types of resources available to you, so you know where to look when you need them.

- Chapter 7, "Careers for Math Majors," contains interviews with former math majors who have gone on to a wide range of different careers. Even if you don't read every single interview right now, we recommend skimming through the section headers to see what kinds of jobs people with math degrees are doing and then dive deeper on any careers that seem interesting to you.

0.1.2 For students who are already math majors. Now that you have taken most of your introductory math courses, you might be thinking about how to get the most out of your last year or two of college and/or what you will do after graduation.[1] Here are some sections of the book that might help you think about what you will do next based on your interests and experience.

- Chapter 1, "Introduction: Start Here," is worth revisiting frequently. It is likely that your goals, interests, and priorities will change throughout your career. How do you need to change your plan and priorities to meet your new goals?

- Even after you have taken your introductory math courses and declared your major, Chapter 2, "Planning Your Course of Study," has information about different electives you might consider taking to help you meet your career goals. (Or if you're grad school-bound, you can identify some subjects you want to learn more about in grad school!)

- Read Chapter 3, "Extracurricular Explorations," more carefully. There are tons of opportunities to get a paid summer job that is related to your math major. It takes time to find these positions and apply for them, but first you need to know about the options that exist.

- Some (highly capable!) students begin struggling in mathematics very early on while others don't encounter their first brush with failure or discouragement until late in their undergraduate studies or even in grad school. When you hit a roadblock, we encourage you to read Chapter 4, "Failure and Growth," to gain some perspective.

---

[1]By the way, it's absolutely fine if you don't know what you want to do after graduation; each of us, at some point, has awkwardly changed the subject when faced with a question about our future plans from a probing family member.

0.2. For Instructors

- Chapter 5, "Networks and Communities of Support," has information about professional organizations and what to do if you are attending a conference for the first time. If you have a research or internship position, a talk at a conference is a great way to share your work with the community and make connections with people who are interested in your work.

- Read Chapter 6, "Technical skills," more closely at this point. What skills do employers want to see and how can you develop them in your time as an undergraduate?

- Chapter 7, "Careers for Math Majors," contains interviews with former math majors who have gone on to a wide range of careers. As the time to actually get a job gets closer, these interviews will help you get a feel for different career options open to you.

- Chapter 8, "Applying for Jobs," and Chapter 9, "Graduate School in the Mathematical Sciences," are all about two very common career paths for math majors: jobs in industry and graduate school. If you are just starting to think about the question of "what's next?", skimming through these chapters can give you some high-level ideas. If it is the fall of your last year of college and you will be applying for jobs in 3 months, read these more carefully to learn more about how to apply and how to find resources on campus to help your application stand out.

0.2 For Instructors

Many math departments around the country are introducing courses for their incoming math majors as a way to offer their students extra support in navigating their college experience. Such courses help to level the playing field, especially for first-generation college students and others who do not have robust networks of mentors. This book is designed to be a resource for such a course.

If you are teaching an Introduction to the Math Major course, you might implement some activities suggested by the book. For instance, there's a scavenger hunt in Section 3.7 your students could complete to learn more about resources available to them on your campus. There are also several lists peppered throughout the book, like our favorite YouTube channels, which could be used as the basis for a class project. Several other resources are discussed, like *Living Proof*, *Testimonios*, *Mathematically Gifted and Black*, and more. You might consider having your students read a few stories about the experiences of successful mathematicians in these books and websites and write reflections on what they read.[2]

Given how extensive "Part 3: Life After Graduation" is, this book could also be a helpful resource for a senior seminar, populated by students who are planning for what comes after they complete their degree. We explore a wide variety of careers through interviews with former math majors who are in those careers. We also give tips on applying for and interviewing for jobs, as well as applying for graduate school. This book could be a useful supplement to other career-oriented books like the *BIG Jobs Guide* [23] and *101 Careers in Mathematics* [15].

If you're working with students who are further along in their studies, you could consider activities like having each student research a particular career path that

---

[2]https://tinyurl.com/2mxs989d

interests them. Each student could begin by reading the interviews related to their selected career in this book. Then, they could find an alum from your institution or someone working for a local business who is doing a related job and conduct their own interview. They could search for additional information, like job ads and salary stats online, too. Students' investigations could culminate in presentations on what they learned to the class.

There are so many directions you could move in if you would like to teach a class using this book, but we know that not all math departments offer these seminar-style courses. You might still find it useful, as an academic advisor or mentor for students, to have a few copies of this book available for students to check out as needed.

For instance, suppose one of your students is having a crisis of confidence after failing a math exam. After sharing your own words of wisdom, you may point them to Chapter 4 to help them gain some perspective on the situation. As another example, let's say that a student is unsure of which advanced courses to plan in their schedule. You could point them to Chapter 2 to read descriptions of common course offerings that go beyond what they might read in your course catalog. Or if your mentee is trying to learn skills to help them in an undergraduate research project, you could steer them towards Chapter 6. If you have a student who is a member of a historically marginalized group in mathematics, you might point them to Chapter 5 in this book to help them identify communities of support. You could also suggest to your students that they read Chapter 3 to learn more about REUs and internship opportunities.

As you may have gleaned, this book is designed to be a resource in a number of dimensions. Familiarizing yourself with the content by looking over the table of contents and skimming the book will help you connect your students with the resources they need to be successful.

About the Authors

Figure 0.1. Photo courtesy of Carrie Diaz Eaton.

Carrie Diaz Eaton. I imagined a few careers for myself as a child—a math teacher, a geologist, a field scientist. I was heavily involved in math in high school but more captivated by exploring questions about how the world worked. I was sure I was going to be a zoologist by the time I arrived at college, but I found myself taking "just one more math class for fun" a few semesters in a row and helping my dorm colleagues with their calculus homework for fun. As I explored my options in college, I realized I could have the best of all worlds, using mathematical modeling as a method for studying all of the interesting questions I had.

I took courses broadly in science, programming, and mathematics and enjoyed the flexibility of a liberal arts mathematics degree at University of Maine. Though the program was small in size, I had an inclusive environment where—even as a queer Latina—I was encouraged to always reach for more. I was involved in many campus clubs related to my identities, such as Wilde-Stein, Latin American Student Organization, and Spanish Language Club. But I was also involved with Pi Mu Epsilon, our mathematics club, and a community service group called Gamma Sigma Sigma. I was able to take advantage of research opportunities in Geographic Information Systems (aka GIS) and computational neuroscience, and I tutored and graded for mathematics courses. All of these experiences helped inform who I wanted to be as a person and as a mathematician. Ultimately, I decided I wanted to teach college math. This was my motivation to pursue a PhD, and it represented a huge leap for me as I would have to leave the state.

Knowing who I was and what I wanted when I looked for the right PhD program was important because I had learned I really needed an interdisciplinary-friendly experience. At the time, I was focused very closely on a particular research area in evolutionary theory. However, at the University of Tennessee, I paid attention to the rise of statistics and programming in mathematical modeling and kept learning new things. While my academic teaching career started at environmental liberal arts colleges, I have moved into more data-oriented research in social justice and in interdisciplinary STEM education in a computational studies department. I have also translated this to professional service on committees for professional societies, planning math institute research programs, and organizing conferences. I could not have predicted 25 years ago as a first-year student the path I am on today, but I know I have relied heavily on what was a foundational undergraduate experience.

My journey was informed by taking courses I enjoyed, by getting involved in campus clubs, by taking advantage of a variety of opportunities, and by trusting myself to understand what motivated me and what skills I could bring to the table. I also benefited from great mentorship from a variety of places as well as communities of support. These are the insights I bring to my own mentoring relationships with my research students and advisees. This book is not a substitute for making these important experiences and relationships, but hopefully it adds another set of viewpoints and suggestions to help guide your own path. Thank you for letting us be part of your journey!

Allison Henrich. When I applied to college, I applied as a theater major. I had been involved in every theatrical performance my high school had put on, and this was the absolute best part of high school for me, so it seemed like an obvious choice. In fact, towards the end of my time in high school, I was *only* taking theater classes there. *Wait, what?* Washington State had a program, called "Running Start," that allowed juniors and seniors in high school to take some of their classes at a local community college to earn both high school and college credit. I appreciated the sense of freedom that came with going to Bellevue Community College (now, Bellevue College), not to mention the wider variety of courses I could take.

Figure 0.2. Photo courtesy of Yosef Kalinko.

At BCC, I took calculus, but I also took courses in linguistics, sociology, and first aid. One type of course I was particularly drawn to was philosophy. I loved to think and discuss, to read and write. So, I enrolled in an ethics course and in Introduction to Logic, and before long, I was hooked. My plans shifted. Instead of pursuing theater, I resolved to declare a philosophy major when I transferred to the University of Washington to complete my bachelor's degree.

In the philosophy department at the UW, I kept on going with logic, taking Advanced Logic, Axiomatic Set Theory, and Computability Theory. Through these classes, I became friends with a delightful crowd of logic geeks. Everyone was majoring in philosophy with a second major or a minor in fields like math, computer science, and microbiology. So, it felt natural for me to return to math and explore a bit more.

It started innocently enough with the calculus sequence, but then I found myself taking courses like Linear Algebra and Differential Equations. When my Differential Equations professor found out I wasn't a math major, he was totally confused. He encouraged me to declare a second major in math. And so, I did. The further I got in my studies, the more fascinated I became with this beautiful subject. And the math department is where I found invaluable mentors who helped me find my path to grad school.

So much has happened since then. I earned a PhD in math from Dartmouth College. I became a professor at Oberlin College for a moment, and then I shifted to a career as a professor at Seattle University. Over the years, I've mentored dozens of undergraduates in research, and I've published many more research papers and books than I ever thought possible. Now, part of my time is devoted to my job as editor of *MAA FOCUS*, the news magazine of the Mathematical Association of America. It's been a twisty path, but I wouldn't have it any other way.

Steven Klee. I completed my undergraduate studies at Valparaiso University, a small liberal arts school in Indiana which is perhaps best known for having had a good basketball team in 1997. The town is equally famous for hosting a popcorn-themed parade in honor of Orville Redenbacher every year because Indiana is just that exciting.

Starting during my second year of undergrad, I was able to get involved with undergraduate research projects at Valparaiso. I also participated in an REU at Central Michigan University during the summer after my junior year. These were formative experiences in my mathematical journey, which not only got me excited about playing with unsolved math problems, but also made me appreciate the importance of having a strong network of mathematical peers and mentors, several of whom continue to be valued sources of wisdom 20+ years later.

Figure 0.3. Photo courtesy of Steven Klee.

This excitement led me to pursue a PhD in math at the University of Washington. After a two-year postdoc at the University of California, Davis, I spent nine years as a professor in the Mathematics Department at Seattle University. I then left academia for industry, where I currently work as an applied scientist with Amazon Web Services. I have been fortunate to be able to karmically pay back my professors' generosity by mentoring undergraduate research students (in academia) or summer interns (in industry) almost every summer since 2011.

Another experience which was instrumental in shaping my mathematical life was participating in the Budapest Semesters in Mathematics program. I grew tremendously as a mathematician during my time in Budapest, where I learned to embrace the struggle as part of the mathematical process. I also grew tremendously as a human through my experience as an outsider who did not understand the language or cultural norms.

A final important piece of my undergraduate experience was participating in the Putnam exam for three years. I never had a particularly outstanding score, but I always enjoyed the challenge of thinking about hard problems, even if I couldn't solve most of them. This experience instilled an interest in contest mathematics which has become a major part of my professional life: I have been the co-director of the University of Washington Math Hour Olympiad since 2012 and have served as an associate editor (and now co-editor-in-chief) for the American Mathematics Competition (AMC 8) since 2019.

Figure 0.4. Photo courtesy of Jennifer Townsend.

Jennifer Townsend. When I landed at Scripps College (a small women's college that is part of a consortium of five liberal arts colleges in Claremont, CA), I didn't know what I wanted to major in. The one thing I *thought* I knew was that I wasn't going to major in mathematics: it was the field that, so far, felt like both the least intuitive and the most rigid. But looking at the course catalog, I was drawn to the post-Calculus math courses which described areas of knowledge I hadn't known existed. I signed up for a couple of those courses just to explore. The inspiring professors, proof-based content, and curiosity and intelligence of my peers pushed me to see mathematics as a creative and thought-provoking field. I was hooked. At the time, I didn't think about career choices, just about sating my curiosity.

I was lucky enough to get early direction from professors to participate in varied mathematics programs. These included: Summer Mathematics Program for Women Undergraduates at Carleton College (sadly no longer funded), Budapest Semesters in Mathematics (a joyously intensive immersion in both mathematics and a new culture), and the SMALL REU at Williams College where I was fortunate to have my co-author Allison Henrich as my first mathematics research mentor, which serves as evidence that your connections as an undergraduate can lead to unexpected opportunities decades later! Each opportunity opened doors for a subsequent one.

Although I wanted to go to graduate school, I first took a year off. I quickly realized that my pure mathematics classes didn't translate into finding a job quite as smoothly as I'd expected. Ultimately, it was a connection from a classmate and the handful of computer science classes I'd taken on the side that led to my first job as a software engineer.

I later went on to attend Georgia Institute of Technology, where I filled in gaps in my mathematics education and took a host of statistics and machine learning courses. I left with "only" a master's degree once I got a job offer for a position at Bellevue College. I taught, got tenure, co-developed a CS curriculum, led undergraduate research, and loved my time at Bellevue College for seven years before I transitioned to join the tech industry in 2020.

Now that I'm settled in industry, I'll freely admit that most of the content from my advanced mathematics courses is unused and forgotten. I can also admit that I was naive regarding the prospects for turning a math major into a career. I underestimated the value of adjacent courses like statistics, programming, and (not offered at the time) data science. Gratefully, I've learned that there are many opportunities for filling the gap between a math major and "marketable skills," most easily in undergraduate courses, but also through post-baccalaureate programs or intensive online or self-study. And despite ending up in a different career, I certainly do not regret mathematics as a field of study. Intensive study of mathematics and collaborations with mathematicians re-wired my brain such that I became better at understanding, simplifying, communicating, and solving challenging problems. I would not be the person I am today without having studied mathematics.

I am currently a Principal Data Scientist and manager at Microsoft, where my colleagues and I research and apply statistical methodology for technical applications. This work was created independently of my employment at Microsoft.

Part 1

Exploring Your Interests

1
Introduction: Start Here

> "There were four **P** thoughts that I had...
> - Mathematics is my **passion**. What I want to do is be a mathematician.
> - What classes and experiences would I need to be a mathematician? How will I **prepare** to be a mathematician?
> - If I worked the **plan** that I had—if I found out what the classes are that I need to take and started taking them, that would put me on the road to preparing.
> - And then I would need to **persist** in this journey to be a mathematician, to persist when I would run into problems. I would have to solve those problems and then continue on my journey without quitting.
>
> That's the way I would get to be a mathematician."
>
> -Dr. Christine Darden, MAA MathFest, August 6, 2021[a]
>
> ---
> [a] https://bit.ly/Darden-MathFest

Welcome to your guidebook! We're so happy you're here.

Before you explore all of the resources this book provides, we encourage you to start here with some self-reflection. The first step in any journey should be to check in with yourself. Who are you, what do you value, and what are your passions?

Figure 1.1. Dr. Christine Darden in Langley's Unitary Plan Wind Tunnel in 1975. Credit: By NASA.[1]

Let's begin by channeling the four Ps in Dr. Darden's quote:

(1) Passion: Your passion helps you see yourself as a mathematician. It motivates the work that follows.

(2) Plan: To set yourself on a path for success, plan what classes and experiences you need to achieve your mathematical goals.

(3) Prepare: Invest in your plan! Find the strategies, resources, and support you need to achieve your goals.

(4) Persist: Do the work, even when the math gets tough.

1.1 Identity

Like many words, identity has a specific meaning for mathematicians. But let's consider for a moment the definition that comes from psychology. According to the American Psychological Association, identity is "an individual's sense of self defined by (a) a set of physical, psychological, and interpersonal characteristics that is not wholly shared with any other person and (b) a range of affiliations (e.g., ethnicity) and social roles."

One's identity is multidimensional and can be measured along many axes. As you think about your identity, you might think about gender, race, ethnicity, nationality, politics, sexuality, or religion. But also: are you a listener or a talker? Are you an introvert or an extrovert? The way you approach a problem may also be part of your identity. Do you pride yourself on leaps of intuition? Or on your

---

[1] http://www.nasa.gov/centers/langley/news/researchernews/rn_CDarden.html, Public Domain, https://commons.wikimedia.org/w/index.php?curid=38582453

1.1. Identity

methodical approach? Are you fast, priding yourself on how quickly you can solve a problem, or do you prefer to take your time to study and think?

Some identities become important to you only in certain settings. For instance, did you discover in college that you're part of a group of "first-generation college students" or "nontraditional" students? Sometimes identities are fluid: constantly shaped and redefined by your life experiences. Perhaps you have always thought of yourself as a "math person" because math classes have seemed easier than classes in other subjects. Or perhaps you did not think you were a "math person," but you're now really interested in the subject because you have been in classes that reveal more creative depths or include hands-on or computational ways to engage! All of these interests, experiences, and identities combine to make us who we are. Embrace your uniqueness as a strength!

You might be thinking, "Why is a math book starting by talking about identity?" While writing this book, we spoke with many folks who have found success in their mathematical journeys (and beyond). Each person's journey and definition of success differed according to their unique constraints, passions, and (ultimately) identity. In a book that promises to have lots of advice for math majors, we acknowledge there's variation in the advice and path that works for each person. It's important for you to assess all advice through the lens of your identity and needs. So before you dive into this book, take a moment to think about who you are and what's important to you.

What are some important aspects of your identity outside of mathematics?

(1) _____

(2) _____

(3) _____

What are some important aspects of how you identify that relate to your study of mathematics?

(1) _____

(2) _____

(3) _____

PS: If you identify as a "hardcore" math person, taking every math class you can—that's great! And if you take the classes you need to get a math degree while you take classes for a second major or various interests, that's great too!

1.2 Values: What's Important to You?

If you are studying mathematics, we expect that you value some aspect(s) of the abstraction of patterns that is mathematics. Perhaps the intellectual challenge appeals to you. Maybe you're drawn in by applications to other fields or beautiful logical arguments. You might be interested in studying math because you enjoy being a member of the community of math majors. No matter which of these reasons resonate with you, understanding what aspects of mathematics you value is important. Your values will shape the career options you consider, whether (and where) you go to graduate school, and more.

You have values outside of mathematics that should also influence your mathematical journey, adding constraints such as geographic location. Is it important for you to remain geographically close to your family? Do you practice a religion that informs your lifestyle? Do you value time and opportunity for a personal interest or hobby? Is a regular routine important to you? Is variety or novelty? Do you value setting your own goals and your own approach? Do you value well-defined structure and goals? Do you value collaborations and social connections with colleagues?

There is no Platonic form of a "successful mathematician." Success won't look the same for everyone. To be a "successful mathematician" is something that you define for yourself. **You** define your own success. Some mathematicians are passionate about proving new theorems while others want to explore math problems related to biology or chemistry. Others are passionate about teaching or community outreach and social justice. Math is flexible. You can study math for math's sake, or you can use math to understand something you are passionate about. All you need to do to be a "math person" is (1) be a person and (2) be curious about math or its uses.

To learn about the paths that other math people have taken, we recommend that you read the book *Living Proof: Stories of Resilience Along the Mathematical Journey* [20], which can be downloaded for free from the Mathematical Association of America's website[2] or the American Mathematical Society's website.[3] *Living Proof* contains stories written by 41 mathematicians about times when they struggled with math and how they persevered to find their place in the mathematical community. In these stories, we learn from mathematicians who became "successful mathematicians" in different ways on their own terms. We will refer to various stories from *Living Proof* throughout this book.

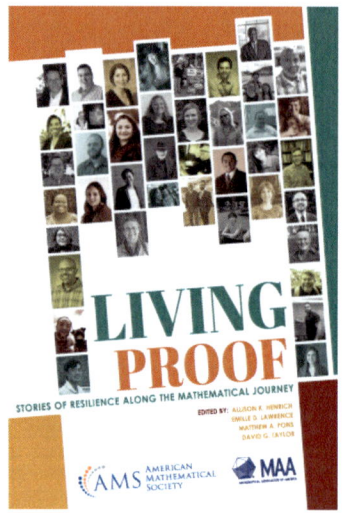

Figure 1.2. *Living Proof: Stories of Resilience Along the Mathematical Journey.* Image courtesy of the AMS.

Later, we'll discuss the importance of mentors and role models as examples of how those who have gone before you have navigated their educational waters. But your path will be unique to you and your interests. What do *your* passions and

---

[2]https://www.maa.org/livingproof
[3]https://bookstore.ams.org/lvngproof/

past experiences make you excited to study? Sometimes it is hard to see how your interests could be related to a math problem, but often your professors and mentors can!

For example, one of my (CDE) students had a family member struggling with Lyme Disease. She was a wildlife major and really enjoyed field work as well as teaching mathematics. We started a research project on tick ecology, which led to an internship, and then to an offer of funding to join a PhD program! Now she's Dr. Alexis White and works as a New York County epidemiologist.

As another example, I (SRK) had a student who played on our university's basketball team and was interested in taking a more analytic approach to understanding the team's performance. He was able to use his expertise as a basketball player to develop insightful models that he could share with his coach. The coach ended up making decisions based on his results, and the student was accepted to a prestigious Master's in Data Science program after graduation.

What is important to you? What are your passions or interests that compel you to action?

(1) _____

(2) _____

(3) _____

Let's put this together! Make a visual representation of your identities, strengths, and passions. Get creative! If you gravitate towards the tactile, grab some poster board, glitter, and pictures from magazines. If you prefer digital design, create a slide show or a vision board. If you're a performer, choreograph a sweet dance number and shoot a video expressing your passions.[4]

1.3 Constraints and Obligations

"Constraints" and "obligations" may be negative words to use to express aspects of your life that are important to you—and lists of these aspects may fit better in the section above on identity. Nonetheless, constraints and obligations are important things to consider when making decisions, so don't neglect to consider them even if they don't feel like they define who you are.

---

[4]Nancy Scherich won the Dance Your PhD competition by creating a performance illustrating representations of the braid group! `https://www.youtube.com/watch?v=MASNukczu5A`

Here are some examples of constraints that might affect decisions you make about your education and extracurricular opportunities.

- You want or need to live close to your family.

- You don't have money set aside for your education, so you need to find educational opportunities that come with funding.

- You or a member of your household has a health condition that requires you to have good insurance, access to specialized medical care, and/or flexible scheduling to accommodate health care.

- You are an international student whose opportunities (e.g., NSF funded research and programs) are limited by citizenship and visa requirements.

- You have a full-time job, so you need academic programs and opportunities that fit into your work schedule.

Think about constraints that will affect decisions you make.

What are your constraints and obligations?

(1) _____

(2) _____

(3) _____

1.4 Goals and Interests

At this point, you've thought a bit about who you are, what drives you, and what other factors you need to consider when making practical decisions. Next, let's think about where you want to go. What qualities do you want in a career? What about in life?

To start off, we've written a few examples of priorities you might have. Rank each one with the following codes:

- P1 means this priority is essential, critically-important.

- P2 means you're unlikely to be happy unless this priority is met.

- P3 means that if these don't happen you'll be sad, but it'll be ok.

- Cross out any priorities that you feel are irrelevant for you.

There are a few empty lines to add some additional priorities at the end of the list. We also encourage you (especially for any P1 and P2 items) to write out specifically what you are looking for in this area.

1.4. Goals and Interests

How critical is...

(1) _____ being able to decide the geographical area in which I live?

(2) _____ making a positive impact on broader society?

(3) _____ having a positive impact on my local community?

(4) _____ making a positive contribution to my family?

(5) _____ making contributions to knowledge in my field through research?

(6) _____ having a career that allows me to continually grow over the course of my lifetime?

(7) _____ having time to spend with family and/or friends?

(8) _____ having time to dedicate to a personal interest outside of work?

(9) _____ having a clear separation between work and home life?

(10) _____ having external deadlines and/or pressure to stay motivated?

(11) _____ having regular interaction with my colleagues?

(12) _____ being able to decide which types of mathematics I research/teach/apply?

(13) _____ _____

(14) _____ _____

(15) _____ _____

Some of the answers to these questions and others we've asked in this chapter will change over time. We encourage you to take some time to reflect and write about your current goals and dreams. If you have trouble getting started, try free-writing: write whatever comes to your mind—even if it's

i am so sleepy.
i am so sleepy.
i am so sleepy.
i am so sleepy.

until you can get your thoughts flowing.

1.5 Roadmap for the Rest of the Book

This book is divided into three main parts. Part 1 focuses on the academic side of your undergraduate experience. In Chapter 2, we discuss some common paths through different types of math majors. What is the difference between a major in theoretical math, applied math, statistics, or data science? What classes are you likely to take, and what other classes should you take during your first two years of college to help prepare you for the rest of your undergraduate career (and beyond)?

What you learn in the classroom will only be a fraction of what you learn in college. Chapter 3 focuses on extracurricular opportunities where you can apply what you are learning to your passions. For example, we discuss undergraduate research experiences, summer internships, math contests, and ways to get involved in your department, university, or local community.

Math can be hard. So can life in general. Let's just be honest about that. The second part of the book is all about setting yourself up with support systems to help you when math gets hard. Chapter 4 offers advice about what to do when you stumble, fall down, or fail at some point in your mathematical journey. A common theme in Chapters 4 and 5 is one of finding community and networks of support to help you along the way. Chapter 5 focuses more on professional societies and specialized groups that may be related to different aspects of your identity.

Finally, Chapter 6 is all about your transition from college to the work force and the skills you will develop in your time as a mathematician, such as collaboration skills, writing skills, communication skills, and programming skills.

The final part of the book is aptly titled "What comes next?" The answer to this question will be informed by your 4 P's, your strengths, and your goals. Maybe you don't know what you want to do next. That's ok. A lot of people don't know what they want to do, and many will change their paths at some point along the way. In my (SRK) life, I have thought I wanted to be a professional baseball player, a high school math teacher, a civil engineer, an electrical engineer, a mathematician, and a data scientist. I changed my major from engineering to math when I was an undergraduate. I changed my career from academia to industry 20 years after that. I was never very good at playing baseball, but I learned the fundamentals of machine learning and data science by investigating baseball sabermetrics.

Each person's journey through mathematics is unique, and our (the authors') experiences are still limited in the grand scheme of things. In Chapter 7, we share interviews with dozens of former math majors who have gone on to a wide range of careers. Some of these careers are things you might expect: actuaries, software engineers, and data analysts. Others might be more surprising and represent the

1.5. Roadmap for the Rest of the Book

true diversity of careers former math majors end up in: veterinarians, policy advocates, artists, and writers. Our hope is that these stories can give you a sense for the options that may be available after you graduate with a math degree.

Finally, Chapters 8 and 9 focus on getting a job after college. Chapter 8 is all about applying for non-academic jobs, while Chapter 9 focuses on applying for graduate school in math or math-adjacent fields.

Our goal in writing this book is to share things your professors know about navigating your undergraduate journey as a mathematician but that you may not know to ask about, and perhaps a bit of context from outside academia that your professors may not be familiar with. There is no one-size-fits-all math major, and there is no single piece of advice that will work for everyone. You certainly don't need to do everything in this book in order to be a successful mathematician. Rather, we want to tell you about some options so you can get the most out of your time in college and be prepared for whatever comes next. Good luck. Have fun. We're rooting for you!

2
Planning Your Course of Study

In this chapter, we discuss some common areas of study for mathy folks. We will tell you a little about each area, some types of job opportunities that are associated with each one, and what kinds of classes you might take. This is very much an exploratory approach to introduce you to the diversity of options within mathematics.

We introduce a wide variety of options here with the caveat that, depending on your institution, some of these options may not be available. Know that, even if your college or university doesn't offer certain classes or majors, that doesn't mean it's not an excellent school providing you with a stellar education. And if you are interested in broadening your knowledge base beyond the courses you will take to complete an undergraduate degree at your school, you can do so in graduate school or in a bridge or post-baccalaureate program.

As you progress in your mathematics major, you may revisit this chapter to help you design a course of study for your bachelor's degree. With this in mind, we will talk about different things you might consider as you design your degree: from interdisciplinary options to practical career training. We also include a special section for transfer students or students choosing mathematics later in their studies, with particular strategic recommendations.

2.1 Introductory Mathematics Core

There are several courses that math majors or minors are required to complete before upper-level coursework. These generally include algebra and/or precalculus courses, a calculus sequence, linear algebra, differential equations, and a course that introduces mathematical proof techniques. If possible, it's important to complete these courses early, to ensure you have time for upper-level coursework which will generally include a handful of required classes as well as some options you can select from to align with your interests and goals.

We will note that after the interviews in Chapter 7, we noticed a theme in non-teaching careers significantly using data and programming. Therefore, we highly recommend that your core mathematics courses also include a statistics and a programming course, even if they are not explicitly required in your major.

Calculus. While many of us who end up in STEM[1] fields in college are familiar with mathematical techniques (like symbolic manipulation of equations) and concepts (like "what is a function?"), calculus can still be a bit of a shock. Calculus requires a very strong foundation in precalculus and algebra, while adding more abstract definitions and proofs.

The structure of a calculus sequence can differ for different departments, but the core content is fairly standardized no matter what college you attend. There are usually three distinct parts of the calculus sequence:

(1) a course studying how to mathematically represent continuous change of functions (this course is often called Calculus I, or Differential Calculus);

(2) a course introducing how to measure cumulative change of functions (often called Calculus II, or Integral Calculus);

(3) a course addressing how to think about and apply change and accumulation across multiple dimensions (often called Calculus III, Vector Calculus, or Multivariable Calculus).

Why is calculus important to a math major? Ideas from calculus courses are revisited in upper-level courses such as Real Analysis (where you go back and prove nearly everything you took for granted in calculus) and Optimization (what is the largest/smallest value a function can achieve over some set of inputs?). On the applied side, calculus techniques are used in many applied math and science fields, such as physics, probability, economics, engineering, and modeling.

Linear Algebra. Linear Algebra sounds like it should be a class about lines and algebra. This is partly true. It is the study of systems of linear equations (think: lines, planes, and their generalizations to higher dimensions). In the applied side of Linear Algebra, you'll learn how to reason about systems of linear equations geometrically and solve them algebraically using mathematical objects called matrices. This initial content in Linear Algebra is deceptively simple: the practical techniques and theory have impressively wide utility across mathematics (both theoretical and applied), in industry, and in the sciences. To whet your appetite, here are a few topics that you may not see in an introductory linear algebra course but that you might see in more advanced courses that put linear algebra to use:

- linear regression;

- dimensionality reduction, data compression, and key feature extraction (e.g., via the Fourier transform);

- efficient computation methods;

- gaming graphics and CGI;

- recommendation algorithms (e.g., from Google and Netflix), built on matrix factorizations.

---

[1] STEM stands for Science, Technology, Engineering, and Mathematics.

2.1. Introductory Mathematics Core

For more examples, Tim Chartier's *When Life is Linear* [7] provides a number of beautiful linear algebraic vignettes.

Linear algebra has a lot of applications, but many courses on the subject also introduce structures, concepts, and proof techniques that re-appear in theoretical classes like Abstract Algebra and Advanced Linear Algebra.

Thanks to both the direct applications and the abstraction of structure, Linear Algebra is often cited by former math majors as the most directly useful class they took during their undergraduate mathematics coursework.

Differential Equations. Your first exposure to differential equations often comes in an integral calculus course: If you know the rate of change of a function, can you work backwards to figure out what the original function was? What type of function has the property that its derivative at some moment in time is proportional to its value at that moment in time? This is the starting point for a course in differential equations, but now we may have more complicated expressions in one or more equations that relate the derivatives of one or more functions.

Such problems are ubiquitous in applied mathematics, physics, and engineering, where real-world phenomena inform assumptions about the derivatives of some unknown function. A famous example of this is the heat equation, which describes how the temperature of a heated object will change over time. This equation, which is derived from assumptions about conservation of energy and heat conduction, relates derivatives of spatial variables (where are we on the object) and derivatives of time (when are we measuring temperature).

You might also go on to take an upper-level course on partial differential equations, or PDEs, which has the same general idea but now involves equations relating the partial derivatives of various functions. It is fairly common to have a 200-level course that serves as an introduction to differential equations, often taken in conjunction with or just after you take linear algebra, followed by upper-level electives where you'll learn about partial differential equations.

And before you run off thinking that mathematicians know everything there is to know about differential equations, determining whether the Navier-Stokes equations, which are used to model fluid dynamics, always have a unique solution is one of the Clay Mathematics Institute's Millennium Prize Problems.[2] Solving this problem comes with a one million dollar cash prize and, presumably, some amount of mathematical fame and notoriety.

Statistics. As such an applied field, statistics is best described using an example. Consider a political poll: a sample of voters is used to estimate something about the entire voting population, for example, the percentage of the population who will vote for candidate X. Different political polls—even those asking the same question at the same point in time—will produce different estimates, each of which will differ from the true measurement of the entire voting population. (This is called sampling error.) That's why a poll is often presented with a $\pm$ margin of error, which represents a confidence interval. A poll that surveys more voters generally gives a better estimate than a poll with fewer, and so has a smaller margin of error. This phenomenon can be made more precise using a result called the Law of Large Numbers.

---

[2] https://www.claymath.org/millennium-problems/

In a statistics course, you'll learn how and why the uncertainty of many estimations can be quantified (in simple cases, by using the ridiculously powerful Central Limit Theorem—but always by a reference to the distribution of the descriptive statistic). You'll assess whether a given sample provides enough information to draw solid conclusions about the whole population. You'll learn how to improve estimates when there is additional information to use. And you'll gain an understanding of the limitations and appropriate uses of your statistical toolbox.

Once you have the concepts down, you can apply statistical methods to nearly any data set. The tools from statistics can be used whether you're publishing a research paper about a cancer treatment's effectiveness or just want a robust answer about whether Instagram posts get more likes when they contain 3 vs. 4 emojis.

If you prefer theoretical mathematics, it will be good to be prepared: in many undergraduate statistics courses, the emphasis is on the mechanical "what" and the "how" rather than the deeper mathematical reasoning underlying the "why." You get to the "why" in upper-level versions of statistics courses offered by mathematics courses that involve probability and have a prerequisite course in Multivariable Calculus. Ultimately, whether you love the applied emphasis of an introductory statistics course, the concepts you will learn will certainly be useful for you. For most math-related careers that don't require a graduate degree and many that do, a basics statistics class is a top-tier investment.

Introduction to Mathematical Proofs. Another course most math majors take teaches how to write elementary mathematical proofs. Proof writing is of fundamental importance to a majority of fields within the mathematical sciences. The names of such a course may vary, such as "Introduction to Proofs," "Introduction to Abstraction" or "Introduction to Advanced Mathematics." Or if your department does not have a separate course, proof-writing techniques may be embedded in another course, such as Linear Algebra, Discrete Mathematics, or Real Analysis. In addition to learning proof techniques—like proof by contradiction and mathematical induction—students often learn about set theory, logic, properties of numbers, and even the fact that there are different sizes of infinity!

Students who love their Introduction to Proofs class often (1) love knowing *why* things work, (2) enjoy being able to convince others of facts using sound, logical arguments, and (3) are in awe of the intricate patterns they discover when looking more deeply at familiar mathematical objects. Students in this camp tend to enjoy more upper-level theoretical (sometimes called "pure") math classes. However, writing mathematical proofs doesn't excite everyone. If you (1) prefer analyzing data and using computational tools to solve problems, (2) care more about how a tool can be wielded than how it was forged, and (3) have a pragmatic desire to find answers, then you will likely enjoy pursuing mathematics in its more applied or interdisciplinary forms.

Read on to learn more about different areas of mathematical sciences you can study as you advance beyond the foundation of a college math major's education.

2.2 Upper-Level Mathematics Courses

Abstract Algebra. In elementary school we learn the four main mathematical operations: addition, subtraction, multiplication, and division. Most of the math we do until high school and early college revolve around these four operations in

2.2. Upper-Level Mathematics Courses

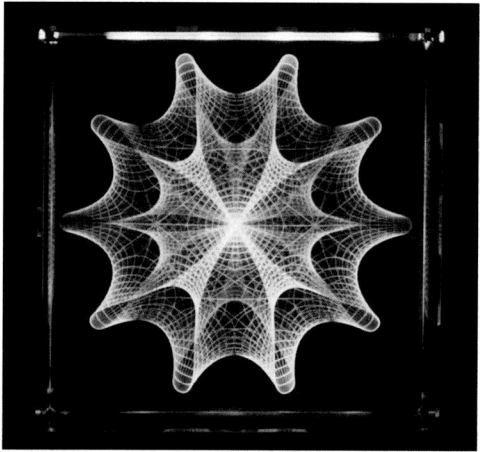

Figure 2.1. The Calabi-Yau manifold in laser etched glass, created by Bathsheba Grossman. https://www.bathsheba.com/crystal/calabiyau/. See Bathsheba's interview in Chapter 7.

one way or another. For example, even in a course on linear algebra, we learn about adding matrices, multiplying matrices and—when it's possible—inverting matrices (which acts like division).

In Abstract Algebra, you will take this idea one step further, working with notions of addition and subtraction and, when possible, multiplication and division, among different sets of objects. The resulting objects are called groups (when you can add and subtract), rings (when you can also multiply), and fields (when you can also multiply *and* divide).

Groups, rings, and fields are interesting to mathematicians because they allow us to axiomatize the requirements of a system where we can "do math" across many areas of application. However, these objects are not merely of interest to theoretical mathematicians.

Symmetry groups are widely applicable in art, where they can be used to encode different types of patterns that appear in tilings or Escher prints. In music theory, groups can be used to encode transformations among notes and chord progressions. Combining group theory with a little probability can be applied to measure how many shuffles are required before a deck of cards become randomized. Group theory is also used in chemistry to study symmetries in molecular structures.

Topology and Geometry. Topology and geometry are two branches of mathematics that both deal with objects like:

- lines, polygons, circles, and knots (1-dimensional objects, often studied as objects in the plane or in 3-space);

- planes, spheres, tori, Möbius bands, and Klein bottles (2-dimensional objects, often viewed as living in 3- or 4-dimensional Euclidean space);

- and higher dimensional spaces—for instance, our familiar Euclidean 3-space, higher-dimensional spheres, solid tori (e.g., doughnuts), and balls (e.g., doughnut holes *yum!*)—but also more exotic objects, like the Calabi-Yau manifold.

There's one big difference between topology and geometry, however. In geometry, objects are rigid. So, we can ask questions about certain physical properties, like angles and curvature. In topology, objects are flexible. They can bend, stretch, and twist without changing their fundamental topological identity. Some artists, like Robert Bosch, have even used this bendy topology to make intricate pieces of art that are topologically equivalent to a simple circle!

Figure 2.2. A sun created by Robert Bosch from a simple closed curve. Learn more about Bosch's art at http://www.dominoartwork.com/.

Real and Complex Analysis. Along with topology and algebra, analysis is one of the three main pillars of theoretical mathematics. Real Analysis often starts by trying to understand the exact reason that calculus works, which means it focuses very deeply on the process of taking limits and becoming good friends with the epsilons and deltas that you may have awkwardly waved hello to in your first semester calculus course. Surprising things can happen and you'll learn that understanding the fine structure of the real number line is critical.

This realization spins off generalizations. What happens in higher dimensions? What happens in spaces with infinitely many dimensions? What happens when you add complex numbers into the mix (spoiler alert: magic is what happens). These questions, in turn, have implications for the solutions of differential equations (both ordinary and partial) which touch upon applications. A year-long sequence of courses in real analysis is nearly essential for students planning a graduate degree in either mathematics or applied mathematics and can also be useful for students studying time series forecasting through the business school. A course in complex analysis is, perhaps surprisingly, essential for students who are interested in applications related to physics or various branches of engineering.

Numerical Analysis. A course in numerical analysis explores methods for achieving approximate solutions to equations through computer simulations. These techniques are important because there are many real-world problems that are modeled by complex systems of equations for which an exact solution either is not known

2.2. Upper-Level Mathematics Courses

(for example, in the case of the Navier-Stokes equations, mentioned previously) or is computationally expensive to compute (it takes too long). In this day of big data and big models, the time scale can be really important.

There are trade-offs to various methods in terms of when they will work, when they won't, how quickly they work, or even whether they will converge to a solution. Sometimes, even if the solution converges, the time it takes to get the solution is impractical or inefficient. In some numerical analysis courses, these ideas are explored primarily through proofs which look at the order of magnitude a particular method will take and/or develop the mathematical theory behind such algorithms. Many of these courses involve programming various numerical methods yourself. How much emphasis is placed on different techniques and skills in Numerical Analysis may depend on your instructor or whether the course serves mathematics majors or also computer science and engineering majors.

Mathematical Modeling. A *model* is "a simplified, abstract or concrete representation of relationships and/or processes in the real world, constructed for some purpose." [8] A mathematical model describes this representation using mathematical notation and language. The modeling process involves (1) identifying what relationship or process we want to describe and the assumptions we will make in order to describe it mathematically; (2) developing, analyzing, and revising the mathematical model; and (3) communicating the resulting insights and findings to others.

Many early courses, such as ordinary differential equations (ODEs) discussed above, may already introduce modeling. However, over the past several years, more and more departments have developed a stand-alone course that introduces how to think like a "mathematical modeler" as well. These modeling courses build out the formalism of the mathematical modeling process and often introduce mathematical techniques beyond ODEs that are used for describing mathematical models. Students are often encouraged to model real-world scenarios in a more open-ended way than in past courses.

Some modeling courses are open to all majors and disciplines with no or few prerequisites. The focus is on a modeling mindset, and they use algebra, discrete methods, and/or computer programs to assist in the analysis. Other modeling courses have ODEs or linear algebra as prerequisites so that they can dive deeper into the analysis techniques.

Often, modeling courses will have themes, perhaps based on the research interests of the instructor. Even if your own interests don't align with the theme of the modeling course, that is okay! It'll still be worth your time. For example, if you are interested in economic modeling, don't shy away from a modeling course that is biology-oriented. Many of the underlying ideas and principles you'll learn in the course are the same as those you'd need to know for economic applications, even if the context is different.

Modeling courses may be offered outside the mathematics department. For example, economics, engineering, physics, biology, and computer science departments often offer such courses. Taking a modeling course from another department can be an opportunity to understand how other disciplines value and use aspects of mathematical modeling. Be prepared: courses outside of the math department often also require a lot more statistical work because of their emphasis on connecting models to real-world data.

Optimization. As its name suggests, problems in optimization are often about finding the optimal solution to a problem. The first time we take a rigorous approach to solving optimization problems is usually in a first-semester calculus class, where we find global/local minima/maxima of functions, possibly subject to some constraints (such as $x \geq 0$ or $-2 \leq x \leq 2$).

In an optimization course, you will learn to study similar problems with one or more possible variations. What if there is more than one variable (maybe you can produce two different items and want to know which combination of them will give you the greatest profit)? What if you need your solution to be an integer (because you can't make $7\frac{3}{4}$ items)? What if the function you are trying to optimize doesn't have a well-defined derivative everywhere? What if it has zillions of local minima/maxima? What if you can't visualize the graph of the function and look for maxima/minima? Oh, and worst of all, what if you need to find a solution quickly (say, before the sun turns into a white dwarf)? How do you decide how to balance time constraints and the quality of an approximate solution?

Solving optimization problems is a huge business, and knowing how to solve them is a surefire way to be marketable in a wide range of industrial careers. Optimization tools are critical in many different fields, including engineering, machine learning, physics, and economics, so this course may be offered by a department other than your friendly neighborhood math/applied math department.

Introduction to Data Science and Machine Learning. An introductory course in data science typically covers basic techniques for working with data to solve real-world problems. The four main steps in solving a data science problem are data preparation, visualization, analysis, and communication, and these are typically the main themes in an introductory course.

In more advanced data science/machine learning (ML) courses, you may focus on different variations on these themes, such as:

- advanced algorithms and modeling techniques (such as neural networks or gradient boosted decision trees),

- best practices in design/user experience (how do you make your data and visualizations meaningful and accessible?),

- particular challenges (such as working with large data sets or problems that must be solved extremely quickly in a streaming setting),

- different problem settings (such as natural language processing or recommendation systems), or

- professional skills (explaining your solution to a room full of people who don't understand any of the math you did but still need to make business decisions based on your analysis).

Combinatorics, Discrete Math, and Graph Theory. Combinatorics is the mathematical art of counting. Problems you'll encounter in Combinatorics will tend to be of the form "In how many ways can you do X with objects from a set Y if you are constrained by Z?" For example, "In how many ways can you walk from the origin to the point (x, y) if you are allowed to only take steps in the directions $(1,0)$, $(0,1)$, or $(1,1)$?" It may seem simple, but there are entire textbooks—even series of graduate textbooks—related to this seemingly simple area of mathematics.

Graph theory is the study of discrete objects called graphs. The graphs you'll study in a graph theory course—as opposed to the graphs you study in Calculus—consist of dots (called vertices or nodes) some of which are connected by lines (called edges). Again, graphs seem simple enough, but there are many problems in graph theory of immense interest to theoretical and applied mathematicians alike. For example, if the nodes represent computers and edges represent ethernet cables between them, you have a graph that models the internet.

Sometimes, courses in combinatorics and graph theory are often combined. Such courses go by many names, but are often built with some subset of the words "Combinatorics," "Discrete Math," and "Graph Theory." These are interesting in their own right, but they can be especially useful for students who are jointly majoring in math and computer science since they form the basis for the objects that are studied in introductory algorithms and data structures courses in computer science departments.

2.3 Programs of Study

In recent years, we have switched from talking about "mathematics" as a single discipline to viewing the "mathematical sciences" as a diverse ecosystem of interconnected fields and subfields. Some fields that used to be considered subdisciplines of mathematics (like computer science!) have blossomed into their own fields of study. The 2020 Mathematics Subject Classification[3] (by the American Mathematical Society) has 151 pages of mathematical fields and subfields!

The ways in which programs in the mathematical sciences coexist vary from institution to institution. Some schools have a single department that houses mathematics, applied mathematics, statistics, math education, and computer science, while others have split these disciplines across several departments. Some students take formal logic classes in a philosophy department or math education classes through the College of Education. Some students learn to code from electrical engineers and learn about dynamical systems from physicists. The moral of the story is that you may need to look outside of the mathematics department to learn about the wide range of courses in mathematical sciences that are available at your institution. We'll talk about various options below.

2.3.1 Pure or theoretical mathematics.
"Pure mathematics" as a term is sometimes improperly interpreted as meaning other branches of mathematics are "imperfect" or "dirtied by application." This term may more accurately represent the historical context of the way humans have constructed mathematics out of (natural) philosophy over and over throughout multiple cultures. In an effort to move away from these types of connotations, we will use "pure math," "theoretical math," or, simply, "math" interchangeably throughout this book.

Theoretical math is focused on building a deep understanding of why and how mathematics works. Courses in this area revisit fundamental ideas that you have studied intuitively throughout your mathematical lives (for example, what is a number?). You might study the constructions of number systems, such as the natural numbers, integers, rational numbers, and real numbers in one course (like Real Analysis or Introduction to Proofs) and then patterns of these numbers in another

---

[3]https://mathscinet.ams.org/mathscinet/msc/msc2020.html

class (like Number Theory). Pure math often attracts those who love a good logical argument and find beauty in the patterns that emerge when mathematics is investigated up close. Common pure math courses that departments offer include Real and Complex Analysis, Topology, Abstract Algebra, Number Theory, Combinatorics, and Analytic Geometry. Many schools require a subset of these courses at the undergraduate or graduate level, typically at least Abstract Algebra and Real Analysis.

Theoretical mathematics is not generally focused on application, but that doesn't mean it cannot be applied. The boundaries between pure and applied mathematics have blurred in recent years. For instance, topology, an inherently visual field, is the study of geometric objects with a focus on how they behave when stretched, bent, twisted, or crumpled (without cutting or gluing!). Topologists study objects like surfaces (such as spheres and tori), knots, spaces that behave like familiar Euclidean space, as well as "exotic" spaces that provide mathematical artists with ample interesting visualization challenges. The tools of topology can be applied in other pure math fields, but they are also applied to studying the underlying structure of data sets in the realm of data analytics. They can help explain DNA replication and can even reveal the mathematical "tricks" that make magic tricks appear magical!

If you are interested in researching pure mathematics because you think it is beautiful, or for the sake of self-directed original knowledge discovery, you might prefer an academic research career (see Chapter 9 on graduate school). If you are not interested or cannot pursue a job in academia you may be considering a job in BIG (Business, Industry, and Government). There *are* research opportunities in such fields: for example, the National Security Agency hires mathematicians to work on cryptography research. You can major in mathematics and pursue this type of career, but be aware that gaining other skills will be necessary to obtain and succeed in BIG careers. Even the more theoretical research in BIG tends to have substantially more use of computation, applied math, and interdisciplinary tools as the research is always aimed to resolve a particular practical problem of interest to the business or government.

Figure 2.3. Knotted DNA shown through an electron microscope. Photo courtesy of Javier Arsuaga and Nancy Crisona.

2.3.2 Applied and interdisciplinary mathematics.
Theoretical mathematicians often find beauty in the elegance of mathematical ideas. Applied mathematicians find beauty in using mathematics to understand and predict the natural phenomena we observe in the world around us.

Applied mathematics focuses on the practical implementation of mathematics and often on using computers to do math and model physical systems. The process of doing math on computers inherently leads to approximation–for example, a computer cannot store all the digits of π (it would be irrational to even try!). Courses in numerical analysis may consider, among other quantitative questions, how to minimize the small round-off errors that compound over time. A course in

2.3. Programs of Study

optimization may ask us how to use mathematical theory to develop faster computing techniques. A course in partial differential equations aims to understand how phenomena change over multiple dimensions.

Some mathematics programs have separate degrees for "mathematics" and "applied mathematics," while others offer specializations or allow students to choose their own adventure. Typically, applied courses of study include Ordinary Differential Equations, Partial Differential Equations, Numerical Analysis, and Mathematical Modeling.

Interdisciplinary mathematics takes this one step further, emphasizing connections to other disciplines of application. The focus is less on the theoretical mathematical questions and instead is driven by questions arising from the applied topic of study. In an interdisciplinary program, you might take applied and statistical modeling courses from professors who have dual appointments or collaborators in other departments. There is often more computational work required, perhaps less formal proof work, and usually there is an explicit connection between the models and data. Example fields include computational biology and data science. (See the section on Statistics below for more information on data science).

Interdisciplinary programs can allow you to combine math with your passions, such as sports analytics (have you seen the movie *Moneyball*?) and data for social justice (check out Data for Black Lives[4] community advocacy work and QSIDE Institute[5] projects). Interdisciplinary degree programs may be housed within a mathematics department, or they might be part of an interdisciplinary center with connections to multiple departments. If your school doesn't offer interdisciplinary degrees, you can design your own interdisciplinary degree by supplementing your math major with a second major or minor from another department.

As an applied or interdisciplinary mathematics major, you would likely have a lot of skills appropriate for BIG careers, but probably few job postings will be looking for "mathematicians." Instead, a posting will have a position for a data scientist, actuary, applied scientist, or analyst. You may also be prepared to apply for jobs in the computer science field. At the end of this chapter, check out the interview with Lee Johnson, who has a theoretical math degree and a job with NASA!

2.3.3 Statistics and data science.
Statistics is the study of how probabilistic outcomes are realized. When you make an observation about the world and observe a pattern, is that pattern "true" or is it an artifact of the particular data you sampled? For example, if you roll a 6-sided die and it lands on "1" three times in a row, is that explainable just from random chance, or is that an indication that the die is "unfair"? Would the answer change if it happened twenty times in a row? If you find that a combination of medications and lifestyle changes is correlated with higher survival rates of a disease, is the relationship just a fluke of your data set or is there a true connection? How does your model of the relationship change as you incorporate new data points? And how confidently can you use the relationship you identified to predict and optimize survival chances?

Statistics has far-reaching applications—anyone who can collect data (such as physicists, supply chain engineers, video game companies, psychologists, or public

---

[4]https://d4bl.org/
[5]https://qsideinstitute.org/

policy experts) can improve their understanding by robustly measuring the relationships and uncertainty in the data they have collected. Statistical methods are the engine behind data science tools such as machine learning, where models are trained to automatically detect and utilize patterns present in large data sets. For example, when you type emails or text messages, the prompt which suggests a word or phrase to automatically complete your typing is derived from machine learning.

Some schools offer undergraduate degrees in statistics or data science. If yours doesn't, you can supplement any degree with a selection of courses that will help you find a career as a statistician or data scientist. In this case, pay close attention to linear algebra and also take probability, introductory computer science, and as wide a variety of statistics classes as you have available.

If you don't have access to traditional courses in machine learning or computational statistics, a number of online courses are respected in industry.[6] It is very common for job postings in statistics to list a preference or requirement of a graduate degree: a Master's degree generally takes about two years and can be particularly beneficial if you don't realize you enjoy this area of math until later in your undergraduate career.

If looking for a job using statistics in industry, commonly the job title is "data scientist." Careers here generally expect foundational coding skills (for example Python, R, and/or a SQL query language) and a background in statistics. Most data science programs are highly interdisciplinary, so the advice about interdisciplinary programs in the section above is also relevant here.

2.3.4 Actuarial science. Actuaries have been using mathematics to analyze risk and make statistical inferences since before "data science" became all the rage in job postings. There is a significant overlap between the program of study in actuarial science and the one that was outlined in the section on statistics and data science, especially when it comes to the core mathematics courses you will need to take. The key difference comes from the courses that supplement those mathematics courses. Whereas data scientists are likely to take more computational or machine learning courses from a computer science or electrical engineering department, actuaries are typically required to take a good number of courses in economics and finance (typically through the business school) alongside their mathematics, statistics, and introductory computer science courses.

Just as lawyers must pass the bar exam in the state where they want to practice law, an aspiring actuary must pass a series of actuarial exams, which are administered by the Society of Actuaries. Unlike the bar exam, you can become employed as an actuary before passing all of the actuarial exams. In fact, actuaries are required to take a number of exams throughout their careers in order to rise through the ranks. The first two of these exams cover probability and financial mathematics (the cool kids call them the P Exam and the FM Exam, respectively), and it is generally helpful to have passed at least one of them before you start applying for actuarial jobs.

Some universities have dedicated degree programs for math majors who want to specialize in actuarial science. If yours does, then your path is a bit clearer. But if not, don't panic. Even if your department does not have a formal degree in actuarial science, it is likely that they have a recommended set of courses to prepare students for actuarial careers. Your advisor should be able to help you with this.

---

[6] Andrew Ng's Machine Learning class on Coursera has been well-regarded for many years.

2.3. Programs of Study

In either case, the set of courses you take as an undergraduate should be designed with an eye towards preparing you to pass the P and FM Exams.

A summer internship with an actuarial firm can be an important opportunity for actuarial students. These internships typically take place in the summer following your second or third year of college (or both!), and they can be invaluable in helping you get your first job after graduation—in some cases, doing a great job during your internship will lead to a job offer before you even graduate! For more on internships, see Chapter 3.

Finally, there are some geographical considerations that come with actuarial careers. Just as a lot of big tech companies are headquartered on the West Coast, many big actuarial firms are headquartered in the Midwest or on the East Coast. Don't be afraid to search for internships in Chicago, Philadelphia, Boston, or New York, even if your university is not particularly close to these cities.

For more information about what a career as an actuary looks like, check out the Be An Actuary website[7] and interviews with actuaries in Chapter 7.

2.3.5 Secondary education. The paths to a career in teaching mathematics vary widely from one institution to another, and once you graduate, the certification requirements vary from state to state. In general, Secondary Education in Mathematics or Math Education programs are designed to prepare mathematics teachers at the middle or high school level (grades 6 through 12) and lead to certification in the teaching of mathematics.

Depending on the size of your institution, you might not directly major in mathematics, but rather in a program from the College of Arts and Sciences or the College of Education. Most math departments have at least one professor who is either an expert in math education or serves as a liaison between the mathematics department and the college of education. If you are interested in teaching mathematics, it is important to be in communication with your advisor or department chair early in your academic career so that you have a solid plan to complete all your math classes, all your education classes, and any field work requirements (such as student teaching or a teaching practicum).

Math teaching programs provide a foundation in pedagogy and learning theory alongside mathematics courses related to the topics you will be likely to teach in your career. This includes foundational course requirements in Precalculus, Trigonometry, Calculus and, in many cases, courses in Statistics or Probability that are becoming increasingly important pieces of K–12 math education. For upper-level electives, courses in Abstract Algebra, Real Analysis, Geometry, and History of Mathematics are good options for future teachers.

Before you begin your career as a K–12 educator, you will need to be certified. Different states have different certification requirements, but sometimes you earn your certification in the state where your institution is located when you graduate (for example, if you graduate from an education program at a university in Iowa, you might be certified to teach in Iowa when you graduate). Many states use the *Praxis* test as part of the certification process. The *Praxis* website[8] is a good resource for determining state-by-state requirements for teacher certification.

---

[7] https://www.beanactuary.org/
[8] https://www.ets.org/praxis

2.3.6 How to select special topics courses. Many institutions require math majors to complete some number of credits of upper-level mathematics courses on top of the core of advanced courses (such as Abstract Algebra and Real Analysis). This can be an opportunity for you to experience the breadth of advanced mathematics and to find your niche by exploring a range of advanced classes across the spectrum of subjects we have outlined in the Programs of Study section. These classes often provide an opportunity for a professor to teach about an area of mathematics that they are truly passionate about, perhaps related to their research, which means you get to take a deep dive into this area with an expert.

Dear Dr. Xue,

I see that you are scheduled to teach a course on Advanced Dynamical Systems next semester. I am a third-year math major, and so far I have focused my studies on less applied courses. I took Linear Algebra and Differential Equations last year, and I am currently taking Graph Theory and Introduction to Mathematical Reasoning. Next semester, I will need to take Real Analysis and some philosophy courses to round out my gen-ed requirements, and I do not want to spread myself too thin.

Can you tell me a bit about how your course will be structured, what topics will be covered, and what your expectations will be for the course? I think I might be interested in moving in a more applied direction in my mathematical studies, which is part of the reason I am interested in your course. I also heard that you sometimes hire student researchers, so I am curious if this class would help prepare me to potentially work with you on research.

I appreciate your time. I could also stop by your office to chat if it is easier for you. I am available any time after 10 a.m. on Tuesday and Thursday of next week.

Thank you,

Name Namerson

On the other hand, it can be difficult to get a sense of what will be covered in these classes because they are not always part of the standard course offerings. As you are thinking about registering for your courses, you may want to email or talk with the professor who will be teaching the course to find out if it is really something that interests you. In the box, check out an example of an email you might consider sending to a faculty member who is teaching an upper-level topics course.

2.4 Changing Your Path

Even if you are certain about the path you want to take, acknowledge that your priorities and life circumstances will change over time. Re-evaluating your intended career path to adjust to new priorities is a good thing. For example, you may be drawn to the innate beauty of mathematics and consider an academic research career, despite the long road you'll need to travel to earn a PhD and a faculty position. Along the way, you may realize you want to focus more on teaching than research; that you prefer the flexibility, higher income, or direct impact available in industry; or that you have discovered your true passion lies entirely outside of mathematics. What should you do if you are thinking about a change in plans?

(1) Aim to take a breadth of courses that will keep your options open. There are some courses that are beneficial in most mathematical sciences and beyond. These include courses in linear algebra, computer science (introductory through at least a data structures class), probability, and statistics.

(2) Apply to entry-level jobs, even if you feel a little underprepared—most math majors didn't *explicitly* prepare for their ultimate career, but found their quantitative skills a great benefit in diverse careers. If you can get a position related to a new field of interest, don't hesitate. You'll learn a lot on the job and you'll also learn what skills you might need to build up in order to advance in that career into a position you'll enjoy long-term.

(3) Do you need more coursework to land a job in a new career? Consider a Master's degree in a different field. Many people use a Master's degree to switch their field. Masters' degrees can lead to a PhD, teaching jobs (e.g., at a community college), or industry positions. In some fields, like Engineering, a Master's degree helps maximize your earning potential based on the time you spend earning the degree.

2.5 Reflection

Now that you have learned about different degree options and career trajectories, take some time to reflect on your interests. What areas are you considering? What are the next steps?

To learn more about what goals you might like to aim for, see Chapter 7 where we interview math majors who have gone on to a wide range of careers. Here, we give a taste of what you can find in that chapter.

2.6 Lee Johnson: From Green River Community College to NASA

One way we discover what's possible is by hearing others tell their stories. Lee Johnson began his post-secondary education at a community college before transferring to a bachelors degree-granting university where he majored in math. Now, Lee works at the NASA Jet Propulsion Laboratory as a Cybersecurity Engineer. Here, we learn about his journey and hear advice he has for students.

> **Tell us a bit about your undergraduate path, from freshman year to graduation. What did that look like for you?**
>
> **LJ:** I meandered a bit from starting at a community college to graduation at Seattle University. I knew I liked math and computers, but I didn't know enough to guide my coursework. As a result, I took a lot of programming and information technology classes along with math while at community college. In some cases, the classes didn't directly help me graduate, but

Figure 2.4. Lee Johnson. Photo by Theresa Stewart.

that meandering helped me once I made it to Seattle U, where I was able to work with my advisor to put together a plan that I knew I would enjoy and could follow to graduation.

What advice do you have for other transfer students? What should they know/do when they get to the institution where they plan to pursue a bachelor's degree in math?

LJ: I think the best thing to do when coming in as a transfer student is to immediately start building relationships and figure out the resources you have available to you. Learn about the services the school and department have to support you, have conversations with your advisor and other professors, get to know the students that you will be working with, and learn about the opportunities to do research or other ways to engage outside the classroom. Those connections will support you in difficult times and add a lot to your experience, all while building skills and providing experiences that will make you more competitive in the job market after you graduate.

You participated in a summer and an academic year research experience as an undergraduate. What about those experiences were valuable to you?

LJ: Everything about the experiences were valuable to me! In less than two years, I was able to see what research mathematics is and get a taste of what that career path entails, travel to conferences and give poster presentations, meet students and researchers from other institutions, get to know my professors outside the classroom, and even help get a paper published. Those experiences helped me understand what I enjoy and wanted to pursue after graduation while being able to work on interesting math problems.

2.6. Lee Johnson: From Green River Community College to NASA

Tell us a bit about your path after graduation. When did you graduate, what kinds of jobs have you done since then, and where are you now?

LJ: I graduated from Seattle University in 2014. Similar to how I started my undergrad journey, I didn't have a clear goal in mind for my career and was just happy to land that first job out of school. I started by writing data analytics software as a contractor for an aerospace company, and as I got to know the industry and how large companies operate, I was able to make small moves in directions I found interesting. Eventually, that led to writing actual flight software inside the flight computers for another company, and from there I have moved into cybersecurity work for spacecraft and all the systems that support them.

As someone working in industry, what advice do you have for current math majors on how to prepare for and apply to jobs in industry?

LJ: There are two main points I would highlight. The first is that it is very rare for a company to pay people to solve math problems directly, but every company wants employees who can solve problems in a mathematical way (which is your sales pitch to help get the job). For you, this means you should think about what kinds of problems you want to solve and tie your mathematics work to other interests or hobbies. If you enjoy helping or educating people, optimizing the way something works, or building certain things, you can use those interests to find an industry or company you want to work with. When your interests and enthusiasm line up with the company's goals, that will shine through and be an advantage in landing that first job.

The second point is: don't only try the "front door" to get a job. Making a solid résumé and applying to jobs online is definitely a game you have to play, but also attend local meetups, go to conferences, engage people in the communities around your hobbies, email people questions about their job and industry on LinkedIn, talk to your network of other alums, and you might be surprised how easy it is get people to ask you to work for them instead of you asking them for a job. There are orders of magnitude fewer students graduating with a mathematics degree (and STEM degrees more generally) than there are available technical jobs in industry, so companies want to hire you. It is up to you to make yourself visible to those companies so they know you are out there, and applying through their website is only one of many ways to do that.

For more great interviews with former math majors, check out Chapter 7. For more information about research opportunities for undergraduates, see Section 3.1.

3
Extracurricular Explorations

> "Scientific observation has established that education is not what the teacher gives; education is a natural process spontaneously carried out by the human individual and is acquired not by listening to words but by experiences upon the environment."
>
> — Dr. Maria Montessori, *Education for a new world*

In this chapter, we'll talk about opportunities for developing your disciplinary knowledge and honing important skills outside of the classroom. These professional development opportunities could take the form of a paid summer research position at a university, an internship with a company or government organization, or a volunteer opportunity in your local community. They might also involve preparing for and competing in a math contest, either individually or as part of a team of your mathy friends, or finding other ways to be more involved in your math department. Not every enrichment opportunity will make sense for every person. View this chapter as a menu of possibilities rather than a list of requirements: the opportunities you pursue should be based on your interests, your goals, and your passions.

So, what are your interests and goals? Do you plan to become an actuary after graduation? If so, you might try to secure an actuarial summer internship with an insurance company. Are you hoping to go to grad school to study math? Then perhaps you should find a research experience, either with a faculty member at your school during the academic year or as a paid researcher in a summer Research Experiences for Undergraduates (REU) program at another college or university. Are you a student of color looking for a community-based math enrichment experience with folks who look like you? There are paid summer research programs you could apply for. Do you like thinking about challenging problems? Do you like testing your skills and knowledge? If so, you might like to take the Putnam Exam or join a team to compete in the Mathematical Contest in Modeling (MCM).

Extracurricular activities can help you achieve your long-term goals because they give you an opportunity to try new things, to pour your energy into an activity you are passionate about, and/or to demonstrate to a future employer that you have experiences beyond the classroom. This is not to say that you should solely seek out extracurricular activities for the sake of padding your résumé—hopefully, you're engaging in these opportunities because they genuinely interest you—but you should also be sure to add them to your résumé when it comes time to apply for positions in the next stage of your career.

3.1 Research Experiences

> We interviewed former math major Ranjani Sundaresan, who had this to say about her undergraduate research experience:
>
> "The most important lesson I took away from doing research is that a failure is a redirection, more than it is a mistake. Before I did research, I saw answers to math problems as being 'wrong' or 'right.' By doing research, however, I learned that you can use a wrong answer to get into the right frame of thinking to get to your goal. As a professional data analyst now, I work to develop strategies to campaign for better environmental policies, which requires a similar mindset in order to be successful. It's really easy to get discouraged the first few times you hit a roadblock, but the tenacity that doing math research builds in you is a lifelong skill that I'm proud to have taken with me."
>
> You can read more about Ranjani's path from her math major to her current position as a data analyst with Greenpeace USA in Chapter 7.

Should I do research? There are many reasons to pursue a research opportunity. If your goal is to go to graduate school in math or applied math, having a research experience when you're an undergraduate is an especially good idea. Not only will you learn about a mathematical topic you might not be exposed to in your college curriculum, but you will also develop a number of skills that are important for graduate school. These could include programming in a language you've never programmed in before, playing with examples to uncover patterns, making conjectures that have never been tested before, reading math research papers, writing a math research paper, and giving a talk on a technical subject. Perhaps most importantly, you will learn more about yourself during your research experience. Do you actually like doing research, or do you hate it? Or do you LOVE IT? This self-knowledge is priceless as you map out your plans for the future.

You might say, "I don't want to go to grad school, but I still want to learn all those skills you talked about!" This is a great motivation for pursuing a research opportunity! Maybe you are leaning more towards working in industry, in the nonprofit sector, or for a government agency. You'll *still* need to know how to communicate complex technical ideas, approach problems that have never been solved before, and there's a decent chance you'll need some coding skills. (See Lee Johnson's interview at the end of Chapter 2.)

3.1. Research Experiences

Many students we've advised have heard about research experiences but don't get involved for any number of reasons: "I don't think I'm cut out for math research"; "I need to make money in the summer!"; "I don't have time to do research during the school year—I need to focus on getting enough credits to graduate!" The thing you might not know about summer research experiences, like REUs, is that many *are* paid positions. Student researchers are usually provided with housing as well as a stipend that is on par with what you'd earn working a typical summer job. If you're interested in doing research during the academic year, you may be able to earn a couple of course credits each term for your work. In many departments, these credits can satisfy elective requirements for the major. If this is something you're interested in, talk to a professor you'd like to work with, an academic advisor, or the department chair to find out about your options.

What research opportunities are there for undergraduates? Now, you're sold on the idea of doing research. How do you go about finding an opportunity for yourself? First, you can start close to home. Have you heard from any of your friends about professors in your department who do research with students? Has one of your professors given a talk or made an announcement in class about a research opportunity they're offering? Ask around! Maybe there's a professor in your department who you'd work well with and who is looking for undergraduate research students. This professor can tell you about the types of research problems they'd like to work on with students and can tell you if it's possible to get credit or pay for your work. Some institutions have funds that faculty and students can apply for to get paid for working on a research project together. Find out if that's an option at your institution.

If you have trouble finding a research experience at your institution, or if you're excited about the idea of doing research full time as a job in the summer somewhere else, a Research Experience for Undergraduates (REU) program might be just the thing for you. There are a wide variety of REU programs, with each one having its own focus, both in terms of what types of students it's looking for and what types of research projects it offers. There is an applied math REU specifically for deaf and hard of hearing students that has run for years at the Rochester Institute of Technology. The MSRI-UP program at the Simons Laufer Mathematical Sciences Institute (SLMath), formerly the Mathematical Sciences Research Institute (MSRI), was designed with Black, Hispanic/Latinx, and Native American students in mind. Students at MSRI-UP conduct research in geometry, combinatorics, and computational algebra. There are some REUs that focus on supporting elite math students (for instance, the longstanding SMALL REU at Williams College), and there are some that aim to introduce first-year college students to research (including MathILy-EST at Bryn Mawr College).

Most REU programs rely on funding from the National Science Foundation (NSF) or the National Security Agency (NSA), and once an REU is funded it will typically run for 3–4 years before the organizers will need to reapply for funding. This means a program that exists today may not be around next year, but also that new REUs pop up every year! To see which programs are currently being funded by the NSF, check the National Science Foundation website[1] or search for "NSF math REU." The American Mathematical Society also has a webpage[2] with

---

[1] https://www.nsf.gov/crssprgm/reu/list_result.jsp?unitid=5044
[2] http://www.ams.org/programs/students/emp-reu

Figure 3.1. Students collaborating at the SUMmER REU.

current REU information, and Steve Butler maintains his own list of REUs.[3] If you are a member of an underrepresented minority group, you might also check out the National Association of Mathematicians (NAM) webpage[4] or Math Alliance[5] for more information on opportunities for you.

Research experiences for international students. Because they are funded by the federal government, the grants that support REUs can only pay students who are US citizens or permanent residents. This means that many REUs cannot accept international students, although there are some REUs whose host institutions provide funding for international student participants. There are also colleges and universities that will provide their international students with funding to participate in a summer research experience at another school. So, if you're an international student, you might need to do some extra leg work, contacting REU directors or your academic advisor to learn more about your options.

What if I haven't taken many math classes yet? There are some REUs that are geared towards students who have taken few math courses, including students from community colleges. There are also programs called pre-REUs that aim to provide students with a strong foundation for a future research experience. At the time of this writing, there were pre-REU programs in mathematical fields hosted by the University of Utah, Clemson University, and more. If you're early in your educational journey, Google it! See what you find when you search for "pre-REU math." If your background is limited, an even better place to start might be closer to home. See the advice above for tracking down research opportunities at your own institution.

Applying for summer research experiences. Let's say you've decided to apply for an REU or another summer research program. The first thing to know is that

---

[3]https://sites.google.com/view/mathreu
[4]https://www.nam-math.org/
[5]https://mathalliance.org/math-alliance-partners/affiliates/

3.1. Research Experiences

application deadlines vary quite a bit, though most deadlines are between January and March. For the application, these programs will likely ask you some information about your educational background, like courses you've taken, grades you've earned, your GPA, and sometimes even textbooks you've used in your math courses. They will also ask you for several letters of recommendation and a personal statement that explains why you're applying, what you would bring to their program, and what you'd like to get out of a math research experience. A group of students wrote a great blog post[6] with advice about applying for REUs.

Letters of recommendation. Having strong letters of recommendation from professors who know you well is a key component of a successful REU application.

(1) Ask the professors who know you well to write your letters. If you ask a professor who only has a vague notion that you exist and were a student in their class at one point, then you are likely to get an equally vague, and ultimately unhelpful, letter of recommendation. This might mean that some, but not all of your letters are written by math professors. If you have a great relationship with a physics or philosophy professor, and they can speak to your strengths using concrete examples to back up their praise, a letter from them is going to be far stronger than a letter from your precalculus professor from two years ago who you have never actually had a conversation with.

One caveat: you might not be aware of an impression you've made on one of your professors. I once (AKH) had one of my most impressive and memorable students from a certain class email me to ask for a letter of recommendation. Her email began, "You might not remember me, but I was a student in your calculus class..." I was delighted to be given a chance to help her get into a program she was excited about.

(2) Whenever possible, ask your professors **early** if they'd be willing to write a letter for you. Your professors are incredibly busy people. You are probably only aware of 25–50% of the work that makes up their job. For them to make time in their schedule to write a decent letter of recommendation may take weeks. Asking them for a letter a month (or more) in advance will enable them to give it some thought and write a better letter on your behalf. At a minimum, ask them two weeks in advance.

Another caveat: if you find out about your dream program the week before the application deadline, don't let the short timeline deter you from asking for letters and applying. If time is tight, apologize for the late request, explain the circumstances, and give your professor the opportunity to decline if they simply don't have the time to help.

(3) A couple of times, I (AKH) have been notified that I should submit my letter of recommendation for a student through an automated system *before my student has even asked me if I'd write a letter for them*. Don't let this happen to you. While some professors will be understanding about being asked to write a letter in this way, other professors will be annoyed that you expected them to write a letter without first having a conversation about it. Be sure to politely, humbly, and gratefully ask your letter writers if they would be willing to recommend you *before* you list them as a reference.

---

[6]https://tinyurl.com/mrxhdwec

Figure 3.2. Undergraduate researchers Brian Freeman, Bushra Ibrahim, and Jayme Reed from Millikin University. Photo courtesy of Emily J. Olson.

(4) Try to make your letter writer's job as easy as possible. Send them:
- information about the program(s) you are applying to, along with a draft of your personal statement or a paragraph explaining why you are interested in applying to the program(s) and how the program(s) will help you achieve your long term goals;
- your transcript or relevant information about your academic performance; if there are some courses you did poorly in and there are extenuating circumstances you'd like to explain (like a health issue or death of a family member), do so briefly—your letter writer might be able to help you put your academic record in context if it's needed;
- information about how they will submit their letter (is there an online system? will they receive an automated email from a website? do they need to mail an actual letter to an actual person? etc.);
- a list of your relevant academic and non-academic accomplishments; and
- an indication of why you chose this person to write a letter for you. Was there some experience you had in their class that you were hoping they would write about? What strengths of yours have they witnessed first hand?

(5) On the same theme of making your letter writer's job easier, plan to send them reminders about letter due dates as the deadlines approach.

How to prepare a personal statement.
Subtitle: Advice from someone who read over 400 REU applications in one week

I've been sitting in the same chair in the library reading personal statements and letters of recommendation for the past three days. I have a dull pain behind my eyes from hours spent staring at my computer screen. My back hurts. My neck hurts. I am trying to get enough coffee in my system to keep my eyes open.

This is the reality of reading REU applications. Most applications are mostly the same: the applicants are good students who have earned good grades in most of their classes; they all love math and have for a long time; they all want to participate in our REU because they think it would be a great opportunity to learn more about math. And not one of those people got hired. Why? Because they were all the same!

The same principle applies when you're applying for any job, whether it is a summer internship, an REU opportunity, or your first job out of college. There is generally no shortage of qualified candidates, and the people who read your applications are going to see hundreds of applications from people who, on paper, are very similar to you. This means you need to do something to make your application stand out. (Hint: mailing an application with an envelope full of glitter, headshots, and decorative hand soaps is not the answer!)

Be specific about your interests. You are probably applying for a lot of positions. Why are you applying to *this* one? Little details—such as including the name of the school or organization and, if you can find it, the name and title of the person who will be reading your application—go a long way. After you have included these details, say something specific about why you are interested in this position.

Good:

Dear Dr. Cobb,

I am excited to apply for the Applied Informatics REU program at Eastern State University. I did a project exploring migratory patterns of African and European swallows in my data visualization class last quarter, and it made me excited to learn more about the bioinformatics project you will be leading.

Bad:

Dear Sirs,[7]

I am writing to apply for a position in your REU. I am a math major at University College, and I think I can make a strong contribution to your team because I have always loved math and solving problems.

Why are you qualified? Most jobs, internships, or REUs have a list of required qualifications. Don't make the person who is reading the application go on a scavenger hunt to determine whether or not you meet those requirements.

Tell a story. Say something personal as it pertains to your mathematical journey and, if possible, demonstrate that you have other interpersonal or leadership skills at the same time. Be authentic and honest. For example, instead of speaking in vague generalities, saying "I fell in love with math in high school," tell the story about how you convinced your high school to start a robotics club, how you taught yourself how to program a Raspberry Pi to use light sensors to detect the distance to a target, and (no big deal) how the experience of being the club president taught you how to be a better listener when working as part of a team.

---

[7]Did you gag a little bit when reading this? I know I did when I wrote it. And I certainly did every time I had to read it.

Give the readers a glimpse into who you are, even if it isn't directly related to math. Maybe you held the world record for stacking the highest structure constructed out of toothpicks and jelly beans from 2017–2020. Maybe you can solve a Rubik's cube in 5 seconds while blindfolded. Maybe you train sheep dogs to compete at the national level. Why should you talk about these things in your application? Because they make you stand out. Because they show you can work really hard to succeed at something very difficult. Because it means you have interests beyond mathematics, which can be important for team-building. Your experiences matter.

Don't try to hide missteps. You don't need a 4.0 GPA to get into an REU. When we (AKH and SRK) ran an REU, we had plenty of participants whose GPAs were far from perfect, including students who had low grades in (or even failed) previous math classes. Why did we hire these students? Part of the reason was that they were able to explain why the circumstances in their lives led them to do poorly in those classes and how they have learned from those experiences to become more successful in their current classes.

Sharing these details can be difficult and traumatic, and you certainly don't need to divulge information that you don't want to share. But sharing context for these struggles can keep your application from being dismissed immediately.

For example, imagine an applicant who has taken two years of college classes, earning A's and a B+ in their first-year classes, followed by C's, D's, and F's in the first semester of the second year. What can the person reading the application conclude based on this? Maybe there was something going on in the applicant's life that led to poor grades. Maybe they skated through the first year classes because they had seen a lot of the material in high school, and now they are struggling to adjust to the rigor of college classes. Maybe they aren't really interested in college. Maybe they have poor organizational skills and study habits that have finally caught up with them. Seeing these grades leads to a lot of "what ifs" and "maybes." Absent any information, it is easy to assume the worst.

If this has happened to you, try to provide context for the time you struggled without giving away more information than you feel comfortable sharing. For example, "You can see from my transcript that my grades in the fall semester were quite poor. I had a family emergency that led to significant stressors outside of school and left me with limited time to focus on my studies. The problem has been solved now, and my grades in my current classes are back at the same level as they were during my first year. This experience gave me a new perspective on the struggles people can face without others knowing about them. I previously would have assumed that someone who was doing as poorly as I was in my classes was lazy or unmotivated, but now that I have been on the other side, I have learned to think about and listen to other peoples' stories before jumping to conclusions." Then (and this is important) make sure the people who are writing your letters of recommendation confirm that your performance in your current classes is really back at a high level.

What do you bring to the program? You might ask yourself, "How could *my* participation possibly make the program better?" Consider this. Do you tend to understand concepts in math class quickly, and then use your understanding to help classmates when they are struggling? You might be able to play a crucial role in a group, ensuring that everyone on the team is up to speed so that everyone can contribute to the research. Maybe, instead, it takes you a bit longer to wrap your head around new concepts, and you've developed a skill of asking good, probing

3.1. Research Experiences

questions to help you (and others) deepen your understanding. Every research team needs at least one person who is great at asking questions and helping the team clarify their ideas; express an interest in playing this role.

Our Top 7 Favorite YouTube Channels

(1) 3Blue1Brown
https://www.youtube.com/@3blue1brown
Visual, intuitive explanations of complicated mathematical ideas.

(2) Numberphile
https://www.youtube.com/@numberphile
Educational math videos and interviews with famous mathematicians.

(3) Vi Hart
https://www.youtube.com/@Vihart
Mathy doodles, art, and music.

(4) James Tanton
https://www.youtube.com/user/DrJamesTanton
Beautiful math ideas that everyone should know!

(5) Tibees
https://www.youtube.com/@tibees
Math and physics exams famous people took as kids, explanations of math Bob Ross style, and more.

(6) ProfOmarMath
https://www.youtube.com/c/ProfOmarMath/
Putnam solutions, GRE test prep tips, and beautiful proofs of theorems you may have missed.

(7) Mathematical Visual Proofs
https://www.youtube.com/c/MicroVisualProofs/
Animated proofs without words.

Have you taken specific classes that provide you with a strong background for, and interest in, tackling one of the REU's advertised projects? In this case, consider tailoring your personal statement for this program; discuss your preparation and interest as they relate to the REU's specific projects or goals. Finally, expressing a genuine enthusiasm for math and math research can go a long way. This is something I (AKH) *always* looked for in personal statements because enthusiastic students tend to be both curious and resilient when we encounter a major obstacle in the research (which *will* happen). In short, take stock of your strengths, interests, and background, and take some time to consider how these things might make you an asset to a research team.

On the flip side, you might have many ideas of how the program will benefit you. We encourage you to not simply focus on the obvious benefits (e.g., making you more attractive to a future employer or grad program). A research experience is going to look good on everyone's résumé or CV (and saying so will make your personal statement look like 100 others), but what will the research experience help *you* with on a more personal level? For example, will the experience you'll gain by giving research talks in the program help you to overcome speaking anxiety you've struggled with for years? Mention this as a benefit to you of the research experience! Will it help you decide between going to graduate school, so you can become a professor who does research in math or math education, or pursuing an interest in teaching high school students? Talk about these competing interests; identify how the summer program might provide you with self-knowledge that you can use to make your decision. Are you driven to use your math background to help solve problems that will benefit those in underserved communities? If a certain REU is focused on problems of this sort, explain why you are passionate about these problems! Are you from one of these underserved communities yourself? How does this hit close to home for you?

Speaking as former REU program directors, the students we (AKH and SRK) were most excited to accept into our REU at Seattle University were the students whose personal statements made it clear that our program could have a profound positive impact on their lives. Personal statements that say something along the lines of "a research experience will help me get into grad school" or make equally vague statements such as "I have loved math since I was little" are less compelling than ones that tell a more personal story of the student's journey, including accomplishments AND setbacks and failures along with reflections of the sort mentioned above. The personal statement is an opportunity to tell a story about what makes you YOU. You are more than a list of grades on a transcript.

Proofread and get feedback. For the love of Benjamin Banneker, **proofread** your application materials. Typos and grammatical mistakes can be very distracting to people reviewing your applications and can make you look like someone who is not detail-oriented. It is a good idea to have several people read over your personal statement, including some of your math professors, friends, and professionals from your university's writing or career center. It's easy to read what you meant to write when proofreading a document instead of what is actually written on the page. Reading your letter out loud to yourself can also help you find places where your writing is unclear or overly wordy. If your words are not coming across as you wanted them to (or if you did something silly such as misspelling your own name!), it is better to find out before you apply to your dream REU program or job.

What are the odds of getting in? Now that you are geared up to apply for an REU, we would be remiss if we didn't point out that these programs are competitive. For the REU we (AKH and SRK) directed, we would get anywhere from 250–500 applicants each year for 10 spots. It is typical for most REUs to receive hundreds of applications for about 10 positions.

All of this is to say: if you want to have a summer research experience, cast a wide net, apply to many programs, and be prepared for rejection. Don't let the odds discourage you. You can't get into a program unless you apply!

What to do if you don't get into a summer research program? Let's say you receive rejections from every program you've applied to. First, go read Chapter 4 on dealing with rejection.

Next, at the time of this writing, there was a new virtual REU program, called the Polymath REU,[8] for students who are not able to participate in a traditional, in-person summer research program. The program doesn't pay its participants, but

---

[8]https://geometrynyc.wixsite.com/polymathreu

3.1. Research Experiences

it also doesn't cost anything. It's significantly less competitive to get into than a traditional REU, and the deadlines fall after the acceptance deadlines for most other REUs. Student researchers in the Polymath REU are put into groups and assigned a main research mentor, additional mentors, and a research problem. Participants communicate about their project through discussion boards and occasional Zoom meetings. It's far more self-directed than most research experiences, but there is a community aspect of the program and an expert mathematician who can answer questions and help guide the research process. It could be an attractive alternative for some folks.

Keep in mind that research programs come and go, so if at the time you are reading this, the Polymath REU no longer exists, we recommend asking around (ask your professors, directors of REUs you didn't get into, or simply search online) to find out if any similar programs are available.

Finally, in comparison with other sciences, the startup costs for doing math research are quite low. You don't need lab space, chemical reagents, million dollar telescopes, or live animals to start working on a math research project. Of course it is hard to recreate the collaborative and mentoring aspects of a formal research project, but it is possible to learn something new and hone your mathematical thinking skills by investigating a problem informally on your own.

In doing this, it helps to reframe your expectations of a "successful" research experience. Every research project does not need to end with a new theorem or a published paper. Before you were in college or high school or any school at all, you were a child who learned a lot of important things through play. Then, somewhere along the way that curiosity of learning through play went away. This is an opportunity to bring that experience back! Find a math problem to play with and see what you can learn along the way. In fact, the role of play in mathematics is so important that it is one of the five requirements of a complete and fulfilling mathematical experience in Francis Su's *Mathematics for Human Flourishing* [**39**].

Journals published by the Mathematical Association of America, such as *The College Mathematics Journal*, *Mathematics Magazine*, or *Math Horizons* typically have problem and puzzle sections at the end of each issue where you can find a playground with some entertaining math problems to explore. These problems typically have known solutions, but the process of attempting to solve the problem and if you're successful, sending a nice written solution back to the magazine, is good practice.

Another way to start engaging with math research is to find a published paper, read it carefully (for more on reading research papers, see Section 6.3), and work through a lot of examples to make sure you understand the main results in the paper. The playful aspect of this exercise is in coming up with examples that range in complexity. Then think about how you might extend the results to different objects that weren't studied in the paper. If you do this, we recommend getting some advice from a faculty member in your department who can help you find a paper that will be accessible and who can help you think about possible extensions (or who knows about papers that conclude with some unsolved problems).

What to expect from a summer research experience. Let's say you've gotten into an REU. Congratulations!!!!

What happens next? Some programs will place you into a particular research group at the time of acceptance and ask you to do a bit of homework before the

Figure 3.3. A student from the MSRI-UP REU in 2023 presenting on their research. Photo courtesy of Sierra Sutherland / SLMath.

program begins. Others will simply expect you to show up in the summer, ready to learn. In either case, it is likely you'll be given a short crash course in background material for your project and good guidance on how to approach your research problem when you get there. So come prepared to learn and participate, and *please* don't spend energy worrying about your research skills and what you don't already know about your research topic. This knowledge and these skills are things you'll be developing over the course of the summer. And take comfort in the fact that most of your peers are coming into the program nervous about what to expect, just like you.

Speaking of your peers, they will be one of your best resources for your summer research experience, so make a point to get to know them from Day 1. Each person brings their own individual strengths and background, and if you know these strengths, you can get the most out of your team to advance your research project. Outside of working on your project, your peers can be a key part of both your current and your future support system. You will work together and play together—you might do things like cook dinner for each other, go on hikes together, and help each other celebrate milestones in your research, birthdays, or anything else that's worth celebrating. You'll also be there to give each other pep talks when you hit a wall in your research, which *will* happen.

A final word of advice. Doing research comes with its own share of highs and lows. Proving new theorems or uncovering patterns in a data set is exciting, but you also need to deal with a fair amount of failure. Here are a few things you should know about research to help you persevere when you hit that wall (borrowed from [**9**, **10**]).

- We all get stuck and frustrated. When this happens, try the following:
  - Take a break.
  - Explain to someone why you're stuck.
  - Review background material.
  - See if the problem can be modified so it's easier to tackle.
  - Check hypotheses or assumptions.

- Work out a simple example.
- Keep going!
- It's ok to ask, "Why?"

- It's OK if you don't understand an idea the first time (or the second time, or the third time).
- Published work is not always correct.
- It's useful to be open to different ideas and approaches for solving a problem.
- Your project might go in a completely different direction than you expect it to.
- Everything takes longer than you think it will. Be patient.
- It's ok to make mistakes. Making mistakes is a great way to learn!
- Hard work and perseverance are necessary (but not sufficient). In fact, hard work is the most important feature of a successful student.
- You don't need to know (and cannot know) all the background.
- Collaboration is important.
- Research is challenging, but rewarding.

3.2 Summer Internships

Internships are a great opportunity to learn about different types of work in an industrial setting. They will expose you not only to the *type* of work you might end up doing, but also to the *workplace culture* you might experience in an industry (with a caveat that both the type of work and workplace culture can vary wildly from team to team and company to company).

It is unlikely that you will find an internship that is explicitly looking to hire a mathematician, but there are still a number of industries that have internships available in math and math-adjacent fields, particularly if you have applied or interdisciplinary interests. For example: tech companies hire programmers, data scientists and mathematical researchers, financial institutions hire analysts and actuaries, and the government hires statisticians, logicians, and cryptographers.

How do I find an internship? Searching for "analyst" or "actuary" internship positions on job sites like Indeed (or Google job search) identifies internships that many junior or senior math majors would be eligible for. These internships are competitive and generally rarer than "software engineer" internship programs, especially at the undergraduate level.

If you get an internship, treat it as an extended job interview. Many industries host internships explicitly as a way of recruiting talent so that successful interns are often invited back for full-time positions. Even if your internship isn't designed in this way, having an internship on your résumé is a massive boost to your chances of getting your job application seriously considered in related fields.

What are industry/government internships like? Depending on the industry, internships often come with pay. (Some of them pay quite well!) For a summer internship, you will frequently be placed with a team, assigned a mentor/manager for guidance, and given a relatively self-contained project to work on. Some interns work alone or primarily with full-time employees, while others are placed in teams to work on a project alongside other interns. This can vary based on the company and team. Some internship positions are for an entire year and are meant to be completed alongside your coursework. Others run for 8–10 weeks over the summer.

When should I start applying? Some companies start looking for their summer interns in the previous fall, while others hire much later. In general, plan to search for internship applications (at least) twice—once about 10 months before your intended start time and again about seven months before your projected start date.

What are some common internship programs? Some of the most well-known internships are from large tech companies, which are fairly competitive and hire thousands of interns each summer. These tend to be software engineering (or project-management) focused. Many national laboratories and large government organizations also host research internships which are available to undergraduate students. Actuarial and investment firms hire summer interns as well. The Society for Industrial and Applied Mathematics (SIAM) curates a collection of internships in fields adjacent to mathematics.[9]

How do I apply? A job posting for an internship will give you instructions for applying and will likely require at least a résumé and cover letter. The section entitled "How to Write a CV or Résumé" in Section 8.7 and the advice on how to prepare a personal statement from Section 3.1 applies for internships as well.

3.3 Study Abroad Programs

There are a number of programs that provide opportunities to study mathematics outside of your home institution alongside other math enthusiasts. These programs give students opportunities to take advanced courses that may not be offered at their home institutions. They can also balance a desire to have an educational experience outside your own institution with a need to complete major requirements. Notable programs in this category include the Budapest Semesters in Mathematics program[10] and the Penn State MASS program.[11] Both are semester-long programs where you take a range of fast-paced, challenging upper-level (or even graduate-level) courses.

In addition to these larger-scale programs for math majors to study abroad, there may be opportunities for you that are organized by faculty and staff on your own campus. For instance, Professor John Carter at Seattle University takes students in his "Mathematical Models of Near-Shore Phenomena" course to Chile over spring break to meet with water wave researchers and explore the region. Also, students from the University of Richmond recently explored mathematics in Sweden with their professor Della Dumbaugh during a summer study abroad experience [**12**]. To find out more, ask your department chair or your campus study abroad office about possible opportunities.

There is one caveat we feel obligated to note when discussing programs in different geographical areas. Non-white, non-heterosexual individuals do not categorically have the same kinds of experiences in programs abroad due to geographic-specific variation in racism, bigotry, and sexism. For example, expressions of LGBTQ+ identities are criminalized in some countries. We recommend researching these locations and programs in full before applying.

---

[9] https://www.siam.org/careers/internships
[10] https://www.budapestsemesters.com/
[11] https://science.psu.edu/math/mass/

3.4. Math Contests

Figure 3.4. University of Richmond students in a study abroad program in Stockholm, Sweden. Photo courtesy of Della Dumbaugh. (left) Seattle University students ready to learn about water wave research in Chile. Photo by John D. Carter. (right)

3.4 Math Contests

In addition to research opportunities and summer internships, there are a number of other opportunities to participate in extracurricular math contests. Two of the biggest such contests are the William Lowell Putnam Mathematics Competition and the Mathematical Contest in Modeling.

Even though these are contests and prizes are awarded, most people choose to participate because they are fun. They present a challenge that is different from much of what you are likely to encounter in your ordinary coursework and an opportunity for you to apply your knowledge to new types of problems. Aside from the promise of a mathematical challenge, they can also give you a chance to get to know other math majors in your department and develop a stronger community with your classmates.

The Putnam Competition. The Putnam Competition is held every year on the first Saturday in December. For some students, this coincides with final exams, which is not ideal. However, it can give you a break from the stresses of your coursework. The competition consists of 12 problems to be solved over the course of six hours, with three hours to solve the first six problems, followed by a break and then three more hours to solve the last six problems.

The problems on the Putnam Competition are very difficult. They do not require advanced mathematics classes—a first- or second-year college student should be able to understand most of the problems on each set. But they do require that you write proofs, and the grading standards for the proofs are extremely high. Each problem is scored on a scale of 10 points, making the maximum possible score 120 points. In a given year, the average score is usually 4 points (out of 120), while the mode is typically 0.

Why should you consider participating in the Putnam Competition? It's fun in a math-is-hard-but-also-rewarding kind of way. The problems are meant to be constructed in a way that you can play with them and try to arrive at a solution. If you think spending six hours solving math problems on a Saturday afternoon sounds like fun, then you'll probably have fun. If you're on the fence about doing it, there's no harm in signing up, giving it a try, and leaving if you aren't having fun. Typically, the opportunity to have lunch with some of your classmates and discuss/commiserate over the morning problem set can be a nice opportunity to meet new people, make new friends, and possibly learn something new.

Math Puzzle and Contest Problem-Solving Resources

(1) Martin Gardner's puzzle books, such as *My Best Mathematical and Logic Puzzles* or *The Scientific American Book of Mathematical Puzzles and Diversions*. For over 25 years, Martin Gardner wrote an article on recreational mathematics in *Scientific American*. The column introduced mathematical ideas through puzzles and games and had a huge impact on thousands of mathematical learners. The next two books are motivated by puzzles in the styling of Martin Gardner.

(2) *Puzzlers' Tribute: A Feast for the Mind* by David Wolfe and Tom Rodgers.

(3) *The Mathemagician and Pied Puzzler: A Collection in Tribute to Martin Gardner*, by Elwyn Berlekamp and Tom Rodgers.

(4) *The William Lowell Putnam Mathematical Competition 2001-2016: Problems, Solutions, and Commentary* by Kiran Kedlaya, Daniel Kane, Jonathan Kane, and Evan O'Dorney; also see *The William Lowell Putnam Competition Archive* https://kskedlaya.org/putnam-archive/. This is a well-maintained repository of Putnam problems and solutions. The solutions can be a bit terse, but everything is there. The book has more robust solutions.

(5) *Mathematical Circles: Russian Experience* by Dmitri Fomin, Sergey Genkin, and Ilia Itenberg. This book is a treasure trove of beautiful mathematical ideas and challenging problems. It is written for people who want to teach math circles to middle- or high-school students, but the problems themselves are great training exercises for the Putnam exam.

(6) Michael Penn's YouTube channel
https://www.youtube.com/c/MichaelPennMath. This channel has a lot of math resources, but in particular a selection of videos devoted to Putnam solutions, especially for problems A1 and B1.

There are a few tips that are worth knowing for the Putnam exam. First, it's practically impossible to solve all 12 problems. If you can solve one problem and earn 9 or 10 points, then you will already have an above-average score. If you can solve two problems, you're probably going to rank near the top 500 (which is a considerable honor). This means you should focus on solving one problem in the morning and one problem in the afternoon.

The standards for proper mathematical exposition on the Putnam exam are very high, so once you have solved a problem, you also need to be sure you write a very clean, rock-solid proof. The difference between scoring a 9 and a 10 on a Putnam problem can come down to a mis-stated quantifier; the difference between scoring a 9 and a 2 can be two mis-stated quantifiers; scores of 3, 4, 5, 6, 7 are almost never given. It's also worth knowing that the problems increase in order of

3.5. Department-Level Engagement

difficulty, so the first two problems in each batch of six are likely to be considerably easier than the rest.

Mathematical contests in modeling. The Mathematical Contest in Modeling (MCM) and Interdisciplinary Contest in Modeling (ICM) are a team-based, multiday contests, typically held in February. Whereas the Putnam Competition is a contest in pure mathematics, the MCM questions focus on mathematical modeling and the ICM questions focus on interdisciplinary fields such as policy and sustainability. Teams select from a packet of several problems, which can generally be described as coming from discrete modeling, continuous modeling, or data science, and spend a weekend developing solutions to the problems and preparing a paper that describes their approach.

Figure 3.5. The 2023 MCM team from Simpson College. From left to right, the people present are Payton Seo, Kalen Stefanick, and Jenna Woodward. Photo courtesy of Payton Seo.

Why should you consider participating in the MCM/ICM? As with the Putnam Competition, it's supposed to be fun. One of the main aspects of the MCM that is appealing to many students is that it is team-based. You get to work together with a group of your peers to develop a solution to a very complex, real-world problem. Another advantage of working with real-world problems is that there is no "right" answer. This is a great opportunity to be creative in applying your knowledge. In many cases, students who participate in the MCM are able to present their work at conferences, such as regional MAA meetings (see Chapter 5 for more on conferences and professional societies). If you'd like to see examples of ICM/MCM papers produced by students at the end of a competition, the results and modeling solutions are posted online.[12] To hear the perspectives of students who participated in the MCM/ICM and scored in the top 3%, check out "Tackling the MCM/ICM" in *Math Horizons*, the undergraduate magazine of the MAA [3].

If you are interested in participating in the Putnam Competition, the MCM or the ICM, the first step is to ask faculty members in your department for more information. Both exams require institutional support from faculty members who will serve as mentors or proctors. Some departments offer training seminars where you can learn more about the specific requirements and expectations of the contests, so you can enter the contest equipped with a useful set of tools for attacking the problems at hand. Both of these contests let you apply your knowledge and learn something new. While they can be extremely challenging, they can also be extremely rewarding, even if you don't win a prize.

3.5 Department-Level Engagement

So far, this chapter has largely been focused on being involved in activities outside your home institution. But there are tons of ways to get involved in your math

---

[12]https://www.contest.comap.com/undergraduate/contests/mcm/previous-contests.php

Figure 3.6. A math+art event at Seattle University where students and faculty created hyperbolic art designed by Chaim Goodman-Strauss.

department as well. Here we will discuss a few common ways that math departments build community, but this list certainly is not exhaustive. Instead, we encourage you to look out for opportunities similar to these and try them out!

Math clubs. Many departments have math clubs that organize math-centric social activities on campus, such as movie nights, game nights, Pi Day celebrations, problem solving clubs, journal reading clubs, or volunteer activities such as tutoring at a local library or school. Other math clubs are associated with a local chapter of Pi Mu Epsilon or Kappa Mu Epsilon, national honor societies for mathematics in the US (see Section 5.2 for more information).

Participating in math club events is a great way to meet other math majors and faculty, which is especially valuable if you are early in your career or a transfer student. And if your department doesn't have a math club, maybe you and some mathy friends can start one!

Department colloquia and seminars. Some departments also have regular colloquia or seminars where internal or external speakers come and give talks. Beyond the benefits of getting to know your peers and professors, colloquia offer the added bonus of getting to learn about some cool new math that is not part of your ordinary coursework. Sometimes colloquia will have a tea/coffee hour before or after the talk, which can be another nice opportunity to network (and even get some free food!). We encourage you to read the section in Section 5.5 on attending math talks at conferences and what to do when you get lost during a math talk. The same advice applies here.

Tutoring. Working as a tutor is a great way to solidify your mathematical knowledge and communication skills. Often we do not truly understand something at a deep level until we can explain it to someone else, so needing to articulate mathematical concepts as part of a tutoring session is a great way for you to learn as well. The ability to communicate technical information, especially to an audience that is not as mathematically knowledgeable as you, is a necessary skill in almost any job, and it is a skill that is best learned through practice.

Most college campuses have some sort of on-campus tutoring center that needs math majors to work with students in courses at or below the calculus level. In addition, people from the local community often contact math departments in search of math majors to tutor their children in math. You can ask your department administrator if they maintain a list of available tutors if you are looking for off-campus tutoring work.

Top 10 Math Club Activities

(1) Throw a Pi Day celebration

(2) Organize a "Pie a Professor" fundraiser (or, combining (1) and (2), consider what it might mean to host a "Pi a Professor" event!)

(3) Host a game night for math majors and faculty

(4) Coordinate a "problem of the week" contest

(5) Organize a jobs panel for math majors

(6) Host a math-themed movie night (See the "Math in the Movies" box at the end of this chapter for ideas!)

(7) Get a group of mathematicians to volunteer in your community

(8) Leave chalk puzzles or math art around town/campus (for inspiration, see Sophia Merow's article [26] on sidewalk math[a]
or check out the Math Around Town videos from #TacomaMath on YouTube)

(9) Make an activity to explore the fold and cut theorem
(see http://erikdemaine.org/foldcut/)

(10) Host an integration bee, Math Jeopardy, or Family Feud event (see "Survey says? Mathematics!" [21])

---

[a]https://www.ams.org/journals/notices/202105/rnoti-p798.pdf

---

3.6 Community Engagement

Many mathematics departments coordinate outreach programs where students and faculty work together to bring mathematics to their local community. There are a lot of good reasons to be engaged in outreach in this way. Maybe you benefited from such programs when you were younger, they had an impact on your educational growth, and now you want to "pay it forward." Or maybe you wish you had had access to such a program when you were younger and want to help create more opportunities for the next generation of students. Perhaps you volunteered working with kids when you were in high school, and you miss that experience. Whatever the reason, sharing math with people in your local community can be a mutually beneficial and extremely rewarding experience.

What are some ways you can be involved in your community? We will share a few opportunities that we know to exist at colleges and universities around the country. But this list certainly is not exhaustive. The best way to learn about community engagement opportunities at your institution is to ask someone. Ask some of your math professors, browse your math department's webpage, or ask the department administrator to help put you in touch with professors doing outreach.

Some departments have partnerships with local schools, libraries, or community centers where math majors can volunteer as tutors or mentors for young students. This is a relatively easy way to be involved in your community, and it is something you can do on your own even if your department does not have a formal partnership. If you are seeking these types of volunteer opportunities on your own, be prepared to be persistent. Keep in mind that community organizations are typically woefully understaffed and overworked, so if you don't hear back after your first request, you may need to reach out again. If sending an email doesn't work, you may need to call them on the phone (we'll pause for a moment so you can clutch your pearls and fan yourself), or even show up on-site and ask to speak with someone. Once you've done this, stay involved—that is the best way to make a contribution!

There is typically some amount of administrative startup cost in terms of your time and energy (for example, most places will run a background check on you). After that you just need to show up at your assigned time, help kids do their math homework, and have fun. In reality, though, tutoring is so much more than helping someone figure out how to solve problems! You have an opportunity to develop impactful relationships with students, who will come to see you as a role model through your weekly meetings. The relationships you build while tutoring are not only beneficial to the students but also help you grow as a mathematician. Being forced to articulate mathematical ideas to someone who is still learning them is good for your learning. It helps you to make deeper connections between different areas of mathematics and solidifies your understanding of the fundamentals.

In addition to tutoring opportunities, some math departments organize math circles for elementary-, middle-, or high school students. *What is a math circle?* A math circle can be many things, but most of them are built around a few common themes. Math circles are meant to expose young students to "real" mathematics. They are less focused on the standard K–12 curriculum and instead aim to share the heart and soul of mathematics with students, introducing logic, proofs, puzzles, and mathematical modeling in a way that is engaging and fun. Getting involved in math circles is a great way to share your love of mathematics with enthusiastic, young students.

Some departments also offer a variation on a math circle for teachers. Instead of working directly with students, the circle engages with their teachers. The goal here is that by providing teachers with an opportunity to explore mathematics in a math circle environment, the impact of the circle will then permeate into the classroom, bringing positive mathematical experiences to even more students. Volunteering with math teachers' circles can be a great opportunity to meet real life teachers, which is valuable if you are considering a career in teaching.

In addition to tutoring and math circle opportunities, some departments also host annual outreach festivals. The purpose of these festivals might be to provide a full day of fun mathematical activities to students in the community. Such festivals typically require many hands and a real "can do" attitude. You may spend an hour stuffing envelopes, printing name tags, or hanging signs around campus, then spend your morning teaching kids how to play with tangrams, and then serve lunch and empty garbage cans before you help someone organize a large, outdoor stochastic dynamical system. If you choose to get involved, be prepared for a fun, chaotic, and exhausting day of mathematical outreach!

The past three outreach opportunities have all been centered around mathematics education in some way. Other opportunities can exist, though they may be less common. Does your campus have a community engagement office? There may be opportunities to connect with community partners, such as nonprofits who do not have the payroll capacity to hire an independent contractor, but who do have mathematical problems that need to be solved. Helping them solve such problems can make a profound impact in your community, while also honing your problem-solving skills and applying the mathematics you're learning in the classroom.

What other opportunities exist? The possibilities are endless! If there is something you are passionate about, such as robotics or computational linguistics, mentoring girls in STEM, working with high school math teams, or broadening mathematical opportunities for children of migrant farm workers, then there may be other communities of support (see Chapter 5) or community programs where you can share these interests in your community. Don't be afraid to ask around, knock on professors' doors, or send emails to community organizations. And if these opportunities do not exist, don't be afraid to ask "why not?" Any number of highly successful community outreach programs began with a simple question: "Why don't we...?" or "What if we...?"

3.7 Scavenger Hunt

In the past two chapters, we have described some common programs of study and extracurricular opportunities. Now, it's time to find out what unique opportunities exist in your department or on your college's campus! If you have a few classmates who might also be interested in learning more about campus resources, have them join you in a friendly competition! How many points can you accumulate in the following Math Classes and Activities Scavenger Hunt?[13]

(1 point each) Meet a professor in your department from whom you haven't yet taken a class. Learn one interesting thing about them.

(1 point each) Meet a fellow math major you didn't already know. Learn one interesting fact about them.

(3 points) Find your department's administrative office. If the department has an administrative assistant, introduce yourself to them.

(3 points) Find out if your department has a lounge or other community space where math majors can gather. If it does, go visit the space.

(3 points) Where can students go to get help with their math homework on campus? Find out about at least one resource.

(5 points) Find out what resources are available to help students plan the math courses they will take. In particular: Will you be assigned an academic advisor? Is there an advising center? Who can YOU meet with to discuss the classes you will take next quarter/semester and your short-/long-term goals?

---

[13]You can call it Math CA$H if you want. ¯_(ツ)_/¯

- (5 points) Find out if your department organizes math-related talks. If it does, attend one.

    +2 bonus points if the talk is given by a student

    +2 points if you are brave enough to raise your hand and ask a question during or after the talk

    +4 points if you attend a second talk; +3 if you attend a third, +2 if you attend a fourth, and +10 if you attend a fifth

- (5 points each) Meet a professor or grad student in some department on campus who does research in (a) a field of theoretical math, (b) a field of applied math, (c) a computational field such as statistics or data science. Ask them to tell you a little bit about their research.

- (3 points) Find out if there are opportunities for undergraduate students to get involved in research in your department/campus.

- (3 points) Meet a professor or staff member who can tell you about how to apply for an internship, REU, or similar summer program.

- (3 points) Meet a professor who participated in an REU or summer research program when they were a student. Better yet, meet one of your classmates who has participated in such a program. Ask them to share a bit about their experience.

- (3 points) Meet a professor or classmate who had a summer internship in a STEM field. Ask them to share a bit about their experience.

- (3 points) Meet a professor or classmate who participated in a study abroad program. Ask them to share a bit about their experience.

- (3 points) Find out if your department has a math club. +100 points if it doesn't have a math club but you start one. +10 points if you become an active member of an existing math club.

- (10 points) Find out what different math major tracks are available in your department/college. (For example, are there different Math and Applied Math degrees?) What are the differences between the tracks? What are the potential career/grad school opportunities that recent graduates of each track have pursued?

- (3 points) Find out if there are opportunities for math majors to work on campus to support the math department (for example, as a tutor or grader).

- (3 points) Find out if there are any professors in your department who engage in some form of community outreach, and learn about volunteer opportunities.

- (5 points) Read a story from *Living Proof: Stories of Resilience Along the Mathematical Journey*.[14] Tell a friend about what you learned in that story.

- (5 points) Visit *Mathematically Gifted and Black* or *Lathisms* (see Section 5.1). Tell a friend about what you learned.

---

[14] https://www.maa.org/livingproof

3.7. Scavenger Hunt

(3 points) Find out if members of your department attend any regional conferences, such as section meetings of the Mathematical Association of America or conferences on undergraduate research. When is the next conference? Is financial support available for students who want to attend?

(25 points) Find a recent issue of *Math Horizons*, *The College Math Journal*, *Math Magazine*, or a similar publication (How do I know what these things are? And what types of publication are similar? You may need to ask around!) Read an article that seems interesting, and then ask a professor or friend to let you tell them about it. Tell them about which parts of the article were interesting and which ones you don't yet understand.

(5 points) Find out if your campus has an annual celebration of undergraduate research. When is it? What are the requirements for presenting? (+50 bonus points if you present)

(5 points) Find out if your campus has programs to support students who want to apply for grants or competitive scholarships, such as the NSF Graduate Research Fellowship or the Goldwater Scholarship.

(3 points) Find out if students from your department participate in the Putnam Competition. (+10 bonus points if you participate)

(3 points) Find out if students from your department participate in the Mathematical Contest in Modeling. (+10 bonus points if you participate)

(10 points) Learn about the upper-level math classes that are offered by your department. Find out which of those classes are required for your major, and think about which ones you would be interested in taking as electives. How often are they offered? If they aren't offered every year, when will you be able to take them?

(5 points) Find your campus career center.

> +1 bonus point: Learn about their services to help you prepare a résumé and/or CV.
>
> +1 bonus point: Learn about their services to help you write a cover letter for a job application.
>
> +1 bonus point: Learn about their services to help you craft a personal statement.
>
> +1 bonus point: Find out if they have services to help you prepare a LinkedIn profile.
>
> +1 bonus point: Find out if they organize a job fair on campus OR if they can point you to job fairs at nearby schools.

(5 points) Make a list of five professors who you would be comfortable asking for a letter of recommendation if you ever needed one.

> **Math in the Movies**
>
> In Spring 2022, Erin Griffin taught a one credit seminar course titled "Math in the Movies" in the Department of Mathematics at Seattle Pacific University. She has shared with us a list of mathy movies and discussion prompts that you could repurpose for an event at your school!
>
> ---
>
> The success of movies like *Good Will Hunting* led to an increase in movies about scientists and, more specifically, about mathematicians. From works of fiction to bio-pics, these movies have done a lot to increase visibility of an oft misunderstood field. This visibility can, however, prove to be a double-edged sword in the way they propagate biases and stereotypes about mathematicians.
>
> **The Movies**
>
> (1) *A Beautiful Mind*
>
> (2) *The Imitation Game*
>
> (3) *The Man Who Knew Infinity*
>
> (4) *Hidden Figures*
>
> (5) *Good Will Hunting*
>
> (6) *Proof*
>
> **Discussion Questions**
>
> (1) Who were the primary mathematicians in the film? Were they real or fictional? If real, how accurately did the movie portray them? (Looking at movie reviews could be helpful here.)
>
> (2) What was the big problem they were trying to solve in the movie? What was the solution to the problem? (i.e., what theorem/invention was created)
>
> (3) What attributes of the mathematician did the movie choose to highlight? How do these feed into our modern understanding of a genius? How does this understanding affect our perception of mathematicians? Of ourselves?
>
> (4) What role did women play in the movie? Consider using tools like the Bechdel Test (see `https://en.wikipedia.org/wiki/Bechdel_test`) to evaluate this. How does this affect our understanding of who can be a mathematician?
>
> (5) What role did race play in the movie? How does this affect our understanding of who can be a mathematician?

Part 2

Supporting Your Success

4
Failure and Growth

> "You may encounter many defeats, but you must not be defeated. In fact, it may be necessary to encounter the defeats, so you can know who you are, what you can rise from, how you can still come out of it."
>
> — Maya Angelou

It may seem strange to talk about failure and growth in the same chapter, but in reality they go hand-in-hand. Growing as a mathematician and learning new, more complicated mathematics are inherently processes of overcoming failure. Take a moment to reflect on this. What is a concept (mathematical or otherwise) that you struggled to understand at first, but that you have mastered now? Can you think of an example from early in your educational journey? How about something from last year? Something from the past month? If you haven't mastered something completely in the past month, what is something you understand better today than you did a month ago?

Adopting a growth mindset is an important tool to wield as you pursue your goals. For example, changing your self-talk from "I can't do this" to "I can't do this...yet" shifts the focus from a static view of education to an ongoing process of learning. At the same time, there are any number of factors in our lives that we cannot directly control (for example, illness or systemic barriers that prevent equal access to education or opportunities). For such issues, we share information about communities of support in Section 4.3 and information on finding community in Chapter 5.

4.1 Overcoming Failure

Studying mathematics requires a certain amount of comfort with failure. Everyone reaches a point where they fail. Everyone. Whether it is failure to solve a homework problem, failure to prove a theorem, or failing an exam, at some point or another it happens to everyone. *Everyone.* A whole section of stories in *Living Proof*[1] focuses on highly successful mathematicians (including Fields Medalist Terence Tao![2]) who overcame failure at one or more points in their lives. Failure, in fact, can ultimately lead to success, because learning from your mistakes leads you to gain more knowledge. Was it Confucius or Chumbawumba who said, "Our greatest glory is not in never falling, but in rising every time we fall"?

Some people encounter failure early in their mathematical careers, perhaps in middle school or high school. Others experience it when they take their first college math course. Others experience it later in college or graduate school. Others, even later.

So what do you do when it happens to you? Here, we will focus on the question of "I just failed a math test for the first time in my life! Now what do I do?" What does it mean to "fail" a test? It can mean a lot of things to different people. Here, we will examine it in the strictest sense of the word—you earned a grade that puts you at risk of not passing the course and needing to repeat it. However, the advice here can apply more broadly to cases where you feel disappointed at not doing as well on the exam as you typically did in previous classes.

First and foremost, **know you aren't alone because it happens to everyone!** This does not define you as a person. You aren't "bad" at math. You are still a human being who is worthy of being loved and respected. One of the best ways to realize that you are not alone is to connect with a community of support. These communities can also help provide you with strategies, resources, and support relevant to your particular context and positionality.

On the other hand, failing an exam should be viewed as a sign that something needs to change in your approach to the class. There are many productive ways to react to failure.

People react to failure in different ways. Some people get mad. Some people cry. Some people go take a nap. Some people exercise, pour energy into a creative outlet, or talk with a friend or family member. You know what you need better than anyone, and you should give yourself space to attend to these needs. Don't try to force yourself to focus on school work if you know you need to go for a run or spend an afternoon binge watching shows on Netflix and eating a pint of ice cream. As a first step to recovery, take a break to deal with the grief.

Once you have a clearer headspace, there are a few important things to do, which do not necessarily need to be done in order: (1) figure out what happened, (2) learn from your mistakes, (3) get mathematical and non-mathematical support, and (4) make a plan for how to better prepare for future tests.

Have an honest conversation with yourself about why you failed the test. College math is harder than high school math. Upper-level college math is harder than lower-level college math. Graduate classes are harder than undergraduate classes. There are spikes in difficulty at every step of the process, and these spikes require changes in study strategies and habits. Maybe in high school, you could do your homework while simultaneously watching TV, listening to music, walking your dog, and perfecting your choux pastry recipe. In college, that probably won't work. College is harder, and your study habits need to adapt to that increase in difficulty. Here are a few questions you can ask yourself:

---

[1] https://maa.org/livingproof

[2] The Fields Medal is one of the highest honors a mathematician can receive, and it has been described as the Nobel Prize of Mathematics.

4.1. Overcoming Failure

Did I really understand the material? An exam is an opportunity to demonstrate that you understand the material at a deep level. Many students will lament to their professors that, "I understand it when you do it, but then I have trouble doing it on my own." Your professors make these things look easy because they have been doing it for a long time. That doesn't mean it's easy! Having trouble doing problems is typically a sign that you could use a bit more practice, perhaps with a coach.

Was I physically and mentally prepared for the exam? Our brains need sleep, water, and healthy food to function properly. Taking a power nap and pounding an energy drink before you take an exam does not carry the same weight as a good night's sleep, proper hydration, and a healthy breakfast. This practice may work once or twice, but it isn't sustainable.

Our Top 6 Online Resources for Course Content

(1) Khan Academy
https://www.khanacademy.org/
Video resources from elementary school math through differential equations and linear algebra.

(2) Wolfram Alpha
https://www.wolframalpha.com/
Computes integrals and derivatives, graphs complex functions, and more.

(3) Paul's Online Notes
https://tutorial.math.lamar.edu/
Worked examples in algebra, calculus, linear algebra, and differential equations.

(4) Dr. Trefor Bazett
https://www.youtube.com/@DrTrefor
Help with discrete math, linear algebra, calculus and differential equations.

(5) The Napkin by Evan Chen
https://web.evanchen.cc/napkin.html
Approachable explanations of ideas from linear algebra through graduate level math.

(6) Desmos
https://www.desmos.com/
Easy-to-use online graphing calculator.

At the same time, sometimes our brains freeze up or go blank during an exam. This happened to me (SRK) during a qualifying exam in graduate school. The pressure was high—if I failed this exam, I would be kicked out of the program. Instead of trying to force my brain to think, I went for a walk. I walked down the hallway and got a drink of water. Then, I went into the bathroom and washed my face. I walked up and down the hallway again, then went back to the test, mentally recentered and ready to think. The math came back, and I was able to finish the exam. Sometimes, a little break like this is all you need. On the other hand, if you frequently find your brain goes blank in the middle of a test, it might be a sign of test anxiety. We also recommend Ken Millett's story "Anxiety Attacked Me, but I Survived" in *Living Proof*.

How can I keep from repeating the same mistakes? Learning mathematics can be likened to building a house. The new material you learn is often supported by prerequisite material from other courses. It is fairly common, especially in lower-level courses, that students struggle more to fill gaps in their prerequisite knowledge than they do in learning the new material. The hardest thing about calculus is precalculus; the hardest thing about precalculus is algebra. In many cases, we don't know that there are gaps in our foundations until we reach the point where they fail us.

It can be helpful to make an appointment with your professor (or a TA or tutor) to go over the exam. They will be able to help you identify the places where you made mistakes and identify whether those mistakes came from conceptual misunderstandings or gaps in knowledge from previous classes. Once you have identified the source of your mistakes, take time to work on fixing them. In some cases, your professor may be able to suggest some practice problems related to these mistakes. Absent that, there is no shortage of online resources, such as Khan Academy, that offer free videos and quizzes to help you review prerequisite material.

Will this take extra work? Yes. Will it require you to do more while also staying on top of the new material you are learning in your classes? Yes. Is it going to be hard? Yes. Is it worth the investment? Yes.

How can I be better prepared for future exams? Time is one of the most common stressors in our lives. How do we accomplish everything on our to-do list (including work, sleep, and having fun) in the finite amount of time we are given each day? For many people, it can be difficult to make progress on *anything* when there are so many things that need to be done. As a result, many tasks (such as homework or studying) get put off until the last minute at the expense of other important tasks (like eating meals that don't feature the words "just add water!" or sleeping).

A well-structured weekly planner can be a key to being better prepared for all aspects of your life. There is a common expectation that you should spend two hours outside of class for each hour you spend in class. Make appointments with yourself (or a study buddy if group accountability serves as a good motivator for your work style) to devote two hours to your math homework for each hour of class time. Spreading this work out over the course of the week will prevent all of your homework from needing to be done at 3 a.m. the day before it is due. It also gives our brains more time to process the new information, leading to better long-term retention.

Is there something else that is preventing me from thriving in my classes?[3] This is a difficult question. There is no way to account for all the demands and sources of stress in any person's life. But perhaps you can identify something fixable that will enable you to be more successful in your classes.

In some cases, the thing that is preventing us from thriving comes from within. When I (SRK) was in eighth grade, I was struggling terribly in my algebra class. Nothing made sense. I couldn't follow what my teacher was saying in class, and as a result I couldn't do any of my homework problems. It wasn't until ninth grade, when I got glasses, that I realized the reason I had been struggling in algebra was,

---

[3]Credit to Drs. Pamela Harris and Aris Winger [**14**].

at least in part, that I couldn't see the board. Everything became easier once I was able to see more clearly.

There are a lot of students who struggle in a similar way to overcome physical or mental hurdles that are preventing them from excelling in the classroom. The transition from high school to college (or from a 2-year college to a 4-year college) is challenging for all students. A 2022 study[4] showed that among all first-year college students who had accommodations in high school to address a learning difficulty (such as ADHD, dyslexia, dyscalculia, or anxiety) or physical handicap, only *one in three* continued to receive similar accommodations in college. One in three. Are you part of the remaining two thirds? Would accommodations help you perform better on exams and quizzes? Most colleges have an office on campus to work with students who are *legally entitled* to such accommodations.

In other cases, perhaps there is something about the timing of the class that is preventing you from thriving. Maybe you work from 6–10 p.m. and are expected to take online quizzes at 8 p.m. or turn in a homework assignment by midnight. Or perhaps you can't attend your professor's office hours because you're a student-athlete and office hours are scheduled at the same time as your practices. In all of these cases, consider asking your professor for some alternate accommodations.

"I have been having a hard time focusing on the quizzes because I work on Tuesday nights and have to take the quiz while I'm on break. Would it be possible for me to take the quiz earlier instead?" or "I can't come to your office hours on Thursdays because I have practice. Would it be possible for us to meet on Friday instead?" It's possible that the answer will be no. But it's also possible that the answer will be yes! Most people don't know about every single thing that's going on in your life, but they may be willing to help you if you explain why a small change on their part would make a big difference for you.

4.2 Mythical Genius (and Why You Shouldn't Care About It)

There is a common lore that suggests mathematical brilliance is a privilege bestowed upon a few (mostly white, mostly male, mostly European, mostly dead) "isolated geniuses"—Newton, Leibniz, Cauchy, Hardy, Ramanujan, Euler, Erdős. The list goes on. The fact of the matter is that most people who are successful mathematicians, *even the aforementioned mostly dead white European dudes*, are/were successful not simply because of some innate talent, but also because they spent a lot of time developing their mathematical skills.

Mathematical aptitude is a skill that can be developed like any other skill, such as rock climbing, yoga, running, baking, painting, ballroom dancing, or making the Kessel run in less than 12 parsecs. You get better at it with practice, and failure and practice go hand in hand. Failure is inherently part of learning. Persistence matters more than innate talent. Talent only gets you so far. After that, it's all hard work. Coming to terms with the fact that failure is part of the process and viewing failure as an opportunity to grow can make it easier to overcome these momentary setbacks. There is plenty of literature to support this. See for example, studies by Carol Dweck [**13**] and Jo Boaler [**6**].

Warning! There is a danger—particularly for women, for Black, Latinx, and Indigenous people, and for other minorities in STEM—of misattributing failure to

---

[4]https://nces.ed.gov/pubsearch/pubsinfo.asp?pubid=2022071

some innate lack of intelligence rather than acknowledging that failure is part of the learning process for *everyone*. This mistaken belief can be held by teachers, mentors, counselors, family members, friends, and others. Even more damagingly, people can hold this mistaken belief about themselves.

Believing (whether consciously or unconsciously) that you are predisposed to be worse at math can, sadly, actually make you perform worse. This phenomenon is called **stereotype threat**, and it's defined as "a socially premised psychological threat that arises when one is in a situation or doing something for which a negative stereotype about one's group applies" [**38**]. There have been scores of studies demonstrating that, if a person is reminded before a test that they are a member of a group that tends to be less successful on those kinds of tests, the reminder itself will cause the person to do worse on the test.

There is good news, though, with regards to stereotype threat. Just knowing about the phenomenon can help inoculate you to its effects. Also, seeking out role models who look like you can be powerfully protective. (See Chapter 5 to learn more about places you might find them.) You know what else can help you not fall prey to stereotype threat? Having support systems and feeling a sense of belonging in the math community.

4.3 Finding Community

The myth of the "isolated genius" can be detrimental to our mathematical journeys. It can lead us to believe that we are meant to study alone, learn alone, do homework alone, and consequently either succeed alone or fail alone.

> "I have learned to ask for help, to admit when I am struggling, and to lean on my professors and peers. Because—you want to know a secret? Everyone struggles. Even when you see them doing well in class, or publishing papers, or succeeding in any way, that success was most probably built on tears and sweat."
>
> — Alicia Prieto-Langarica, in *Living Proof*

Time out. Go read all of Dr. Prieto-Langarica's *Living Proof* story. It's a treasure trove of good advice that is particularly relevant to our discussions here. After you read her story, read the stories by Matthew Pons, Jacqueline Jensen-Vallin, Laura Taalman, and David Taylor, all of which center on the theme of overcoming failure and finding a mathematical community.

While there are still times when it makes sense for people to work independently, the world of STEM is more collaborative now than it has ever been. Employers *want* to hire people who can collaborate and listen, people who can work as part of a team. And teamwork isn't just a skill that you will need if you choose to pursue an industrial career. More research mathematicians (and scientists) are collaborating on research projects now than they were 50 years ago [**5**]. Professors and teachers need to work together to develop their curricula or plan the future of their departments and schools. Analysts, data scientists, and software engineers need to collaborate with economists and business executives. In short, you need to be able to talk to people about the things you're working on, share responsibility for projects, and settle disagreements in a productive manner.

4.3. Finding Community

If we are willing to believe that your time in college will prepare you not only with a technical education, but also with life skills, then we should also view the undergraduate experience as practice for the collaborative work you will need to do when you enter the workforce. By working collaboratively with your classmates, you can find new perspectives on solving problems and share your own perspectives to help other people grow. A rising tide lifts all boats. None of the authors of this book would be where they are today without the study groups and other communities of support they formed in school.

Of course, it can be hard to find good study buddies, and we recognize that it can be doubly difficult for people coming from minority or marginalized communities.[5] At first, it can be a bit awkward to ask someone if they want to study together. It can feel like trying to make new friends on your first day of school. And, like meeting people on the first day of school, you may go through a few awkward iterations until you find a group that is a good fit for your study habits and personality. Regardless, it is worth overcoming a bit of awkwardness for the potential benefit of making some friends and learning together. Your people are out there! And they probably feel just as awkward about it as you do.

In fact, one thing to know is that most people are at least a little fearful about working with others on math because they're worried their weaknesses will be "found out." It's easy to think that your classmates are understanding concepts more easily than you are—that they have more of a natural aptitude towards math. (This is related to the concept of stereotype threat we discussed before.) Having these—often flawed—assumptions about those around you can make you feel like an impostor who is merely pretending to be a highly capable math student. Practically everyone feels like an impostor at one time or another. Every one of the authors of this book felt like impostors at some point in our math journeys. Please know that the feeling is normal, but isolating yourself when you have these feelings is counterproductive. Instead, lean in. Intentionally engage with the material and your peers so that you can all learn together.

How can you go about this? You'll want to find people to study with who'll value your contributions to the group and make contributions of their own. If you are utterly lost in some math course you're taking, your valuable contribution to a study group might be showing up on time and asking lots of questions! If you are mostly getting a grip on course material, your contribution might be explaining what you understand to a study buddy. If you're in this situation, you'll find that explaining things will not only benefit your classmate, but will also help you deepen your own understanding. And please know that if you're on the receiving end of a classmate's explanation, you are not only benefiting yourself by asking for an explanation—you're benefiting your classmate, too. As math educators, we can attest that one of the best ways to help someone learn math ideas is to ask them to explain what they know.

Now, let's say you find yourself in a study group where you're doing all the work for a bunch of people who haven't started the homework yet. *Not cool!* Something needs to change. Setting norms and expectations for how you will work as a group is just as important as the mathematics you will learn together. If you're a Type A Early Riser™ who likes to have homework assignments completed 48 hours before they are due, then you may not want to join a study group that meets at 2 a.m. the

---

[5]In Chapter 5, we will give more information about different mathematical communities where you might seek friends and find mentors.

day the homework is due. If you process information by talking out loud, accepting that 90% of what you say will be wrong, be aware that there are other people who will only be willing to share ideas after they have had time to work on their own for a while. Knowing when to talk and when to listen is a difficult skill to master. But if you find that you're the only one who is talking, then it's probably time to ask someone else to share their ideas.

During my (SRK) first year of college, I took an honors physics course. I failed the first homework assignment and got a D on the first quiz. Not a good start from someone who was used to getting As on everything. I was lamenting this fact to some of my friends, saying I didn't think I should *need* to go to office hours. I was supposed to be smart. Fortunately, they all responded in unison: "Why not? That's what they're there for." So I did. I went to the professor's weekly homework help session. As a result, I learned a lot more and started to make friends in the class. We formed a study group of our own and met every week to discuss homework problems together. At the end of the semester, we all got A's in the class. By working together and getting outside help, we all learned more and succeeded.

4.4 Getting Back to Work

> "I know that things might not always end... perfectly. Or it may take much longer than expected to overcome an obstacle... small-scale failures like failing a quiz or exam are equally disappointing because they chip away at our self-confidence. I've learned to keep in mind that it isn't about the scale of the failure or the length of the struggle. It is about the lessons we learn along the way and the resilience we build up in ourselves. In the end, the struggle is most definitely real, but without struggle, there is no reward."
>
> — Matthew Pons, in *Living Proof*

He's right. Failure, on any scale, damages our self-confidence and our sense of self in significant ways. Do I belong here? Am I cut out for math? Why am I doing this? Is it worth it? It can be hard to get back up when you get knocked down.

Keep in mind that the struggle is part of the process. If math were easy, you wouldn't be taking classes and devoting years (*years!*) of your life to learning it. And learning is an ongoing process. The body of knowledge you amass during your time in college is only a fraction of what you will learn in your professional career. It is the middle of an educational journey that started before you could hold a pencil and will continue long after you have attended your last college class. There will be failures and successes at each step of this journey. Failure is a momentary setback, a sign that you are still learning, not a sign that you are incapable of learning. You've made it this far. Keep going.

5
Networks and Communities of Support

In Chapter 1, we asked you to think about your identities. We considered multiple axes, from social identities to professional identities. It can be really important to find mentors and engage in identity-specific community support beyond your own department and even your institution. Why is it important to have a community of support?

Professional communities can help you get plugged into formal or informal mentoring programs and tell you about professional norms, events of interest, and programs that will help you develop the skills and knowledge you want to gain. If you join a larger community, opportunities such as competitions, conferences, and job positions are announced over mailing lists and through social media channels. Communities of support may also organize specific events related to your concerns, such as a panel discussion on "Balancing motherhood and a career" or a sequence of blog posts on "How to study for an actuary examination." Because communities of support share interests, you can have in-depth conversations around that professional identity. Instead of "How to get into grad school," a society that serves the mathematical biology community might instead have a session called "Navigating graduate school admissions in mathematics and biology departments."

In this chapter, we'll share information about mathematical communities that can provide support for you as you move through your education and into a career. As we discuss these, we also bring up potential opportunities to learn about mathematics, identify mentors, find role models, present work, and network. Finally, we outline some specific things you may need to know if you would like to attend a mathematics conference or workshop. We would also be remiss if we did not mention a new book, "Count Me In: Community and Belonging in Mathematics" [11], which focuses on the importance of community in mathematics and studies many different communities of support that advance individual and entire groups of mathematicians.

5.1 Mathematical Communities: People Create Mathematics

Do you want to broaden your knowledge of who participates in the math community? Are you the only person you know that wants to be a math major? Is it hard to find mathematicians who look like you? Do you feel like you're the only one of your peers struggling with a certain issue (such as test anxiety, depression, or ADHD)? **You're not alone.** There are many others who have been in your shoes. How do we know? Because we have had our own struggles with learning math, mental health issues, and finding our places within the math community. Fortunately, we have discovered ways to share these stories—from the difficult times to the times when we have thrived in our math careers—and find our people.

To learn more about people who do math, there are several community-driven narrative projects that can give you a glimpse into the lives of mathematicians. The book *Living Proof: Stories of Resilience Along the Mathematical Journey* mentioned in Chapter 1 is one example. One of the contributors, Terrence Tao, is a Fields Medalist who tells his story of how a lack of study skills almost led him to fail an exam in grad school. Another author, Autumn Kent, is a transgender mathematician who saw the mathematics community through the lens of someone presenting as a cis-gendered male and, later, through the eyes of a transgender woman. Robin Wilson reflects on how systemic racism nearly pushed him off of his path to becoming a mathematician. Reading these narratives of mathematicians will open your eyes to issues you've never faced as well as journeys you thought you faced alone.

In addition to *Living Proof*, we recommend checking out *Testimonios*,[1] a book (available in both English and Spanish) that shares profiles of Latinx members of the math community. Starting with their childhood and family, authors reflect on how their experiences have shaped them into the mathematicians they are today.

Here are a few more community-based resources for mathematicians who identify with specific groups underrepresented in mathematics:

- Lathisms[2] showcases Latinx and Hispanic mathematicians in the Mathematical Sciences. Pamela Harris, one of Lathisms' founders, also co-hosts the podcast *Mathematically Uncensored* with Aris Winger.

- Mathematically Gifted and Black[3] and Mathematicians of the African Diaspora[4] feature contributions made by Black mathematicians.

- Indigenous Mathematicians[5] showcases mathematicians with Native American, Pacific Islander, Native Hawaiian, and Chamorro roots.

- Sines of Disability[6] is a community of mathematicians, mathematics educators, and activists who are committed to disrupting ableism in mathematics and beyond.

There are a number of other initiatives beyond these that are driven by professional societies and other communities of support—keep reading for more options and opportunities to engage.

---

[1] https://www.maa.org/press/ebooks/testimonios
[2] https://www.lathisms.org/
[3] https://mathematicallygiftedandblack.com
[4] http://www.math.buffalo.edu/mad/index.html
[5] https://www.indigenousmathematicians.org/
[6] https://sinesofdisability.com/

5.2 Professional Societies

Professional societies are nonprofit organizations which rely on membership fees in order to provide support for the professional community. Depending on the size and scale of the organization, they often have dedicated staff who run the organization, while also relying on their membership to volunteer for leadership positions that help direct this work. Many of the mathematical professional societies below have institution-based chapters or clubs that are a perfect place for students to get involved in activities sponsored by the local chapter and those sponsored by the national organization.

Below, we list a selection of professional organizations that serve the mathematics community and have activities of interest to undergraduate students. This is not an exhaustive list. We also divide this into three sections: large comprehensive mathematics societies, specialized subdisciplinary societies, and specialized societies that serve marginalized members of the mathematical community.

5.2.1 Large professional societies.

American Mathematical Society. The largest mathematics professional society is the American Mathematical Society[7] (AMS), which spans all mathematics disciplines and supports multiple career paths in mathematics, particularly research-based ones. Its mission statement reads [1]:

> "A professional society since 1888, we advance research and connect the diverse global mathematical community through publications, meetings and conferences, MathSciNet, professional services, advocacy, and awareness programs."

Many institutions have department memberships which allow their graduate students to have free membership in AMS (and sometimes undergraduates, so check with yours today!).

The AMS also hosts the Joint Mathematics Meetings, the largest mathematics conference in the United States. This week-long conference is held each year in January, and typically draws crowds of thousands of mathematicians from around the world. They feature ample opportunities to give and attend talks, along with a dedicated (massive!) poster session where undergraduate researchers can present their work. In addition, there are social events meant to build community among different groups of mathematicians, such as undergrads, grad students, LGBTQ+ mathematicians, Black mathematicians, women in mathematics, and more.

Mathematical Association of America. The Mathematical Association of America[8] (MAA) is a professional mathematics society that values teaching and learning, community, inclusivity, and communication. From its website [25]:

> "The Mathematical Association of America is the world's largest community of mathematicians, students, and enthusiasts. We further the understanding of our world through mathematics because mathematics drives society and shapes our lives.
>
> The mission of the MAA is to advance the understanding of mathematics and its impact on our world."

---

[7] https://www.ams.org/
[8] https://www.maa.org

Figure 5.1. Conference-goers at MAA MathFest. © Mathematical Association of America, 2023. All rights reserved.

Many institutions have department memberships which allow their undergraduate students to have free memberships in the MAA. Membership includes subscriptions to the MAA magazine for students, *Math Horizons*, and the expository journals *The College Mathematics Journal* and *Mathematics Magazine*, which feature many articles that are accessible for undergraduate math students to read.

In addition, the MAA has a number of "Special Interest Groups" or SIGMAAs which span a variety of more specialist topics such as "Recreational Mathematics" and the "History of Mathematics." Joining a SIGMAA adds a small additional cost to the membership fee, but most activities planned by SIGMAAs are open to everyone. The MAA also has its own set of blogs on Math Values[9] which focus on the values of the organization.

The MAA is also known for hosting several mathematics competitions for students, such as the AMC 8 for middle school students, the AMC 10/12 for high school students, and the Putnam Competition for undergraduates, which was discussed in Section 3.4.

Due to their focus on students, national and regional MAA conferences are great places to learn about different types of mathematical research since talks are designed for a wide range of expertise. MAA meetings are also great places to present undergraduate mathematical research. The national meeting of the MAA is MathFest, which typically takes place in the first week of August. You can present talks or posters as an undergraduate at MathFest. Regionally, the MAA hosts sectional meetings each Fall and/or Spring in each of its 29 geographic sections. Typically, these meetings are easier to attend because they are hosted within driving distance of most colleges/universities in their sections and are concentrated on a Friday and Saturday to limit the number of classes faculty and students need to miss. These meetings tend to be much smaller (with only a few hundred attendees) and place a strong emphasis on creating opportunities for undergraduates to present their work.

American Statistical Association. The American Statistical Association[10] (ASA) has an emphasis on supporting statisticians and statistics education.

"The American Statistical Association is the world's largest community of statisticians, the 'Big Tent for Statistics.' It is the second-oldest, continuously operating professional association in the country. Since it was

---

[9]https://www.mathvalues.org/
[10]https://www.amstat.org/

5.2. Professional Societies

founded in Boston in 1839, the ASA has supported excellence in the development, application, and dissemination of statistical science through meetings, publications, membership services, education, accreditation, and advocacy." [2]

ASA has multiple pathways to engagement in addition to general membership. For subdisciplinary or industry interests, there are both Interest Groups and Sections. Sections have a joining fee and Interest Groups do not. ASA also has regional chapters as well as student chapters and outreach groups. The annual professional meeting for statisticians is the Joint Statistical Meetings, held in August, in collaboration with several other international societies.

Society for Industrial and Applied Mathematics. The Society for Industrial and Applied Mathematics[11] (SIAM) supports applied mathematicians as well as those who want to pursue careers in industry.

> "SIAM fosters the development of applied mathematical and computational methodologies needed in various application areas. Applied mathematics, in partnership with computational science, is essential in solving many real-world problems. Through publications, research and community, the mission of SIAM is to build cooperation between mathematics and the worlds of science and technology." [33]

Regional SIAM groups are called Sections and subdisciplinary groups are referred to as Activity Groups. The Activity Groups sometimes have meetings separate from the SIAM's Annual Meeting. A project of SIAM, called the BIG (Business, Industry, and Government) Math Network,[12] offers an excellent set of resources for students to learn about different career options in industry after graduation.

Pi Mu Epsilon. Pi Mu Epsilon[13] (PME) is an honor society for undergraduate math majors. "Pi Mu Epsilon is dedicated to the promotion of mathematics and recognition of students who successfully pursue mathematical understanding." [28] Some departments have chapters of PME that students can join, which might host social events. PME also publishes a journal for papers written by undergraduates.

Kappa Mu Epsilon. Kappa Mu Epsilon[14] (KME) is another honor society for undergraduate math majors whose goal is "to promote the interest of mathematics among undergraduate students." [22] Some departments have chapters of KME that students can join, which might host social events. KME also publishes a journal, *The Pentagon*, featuring articles written by and for undergraduate students, fun problems, and news relevant to members of the society.

5.2.2 Specialized subdisciplinary societies. This is not an exhaustive list of mathematical sciences specialty communities, but we aim to give a few examples below. In some fields, subdisciplinary societies are important for keeping up individual accreditation. This may be because they oversee the process (e.g., actuarial science exams) or because they provide professional development credits and opportunities for continued certification (K–12 teaching). They can also be important

---

[11] https://www.siam.org/
[12] https://bigmathnetwork.org/
[13] https://pme-math.org/
[14] https://kappamuepsilon.org/

in helping convene a critical mass for conversations that delve deeper into the field through conferences, journals, education, and other programming.

Society of Actuaries. The Society of Actuaries[15] (SOA) has more than 30,000 members. The society manages its professional accreditation standards and exams specific to its field. It also advocates for the profession: "Through education and research, the SOA advances actuaries as leaders in measuring and managing risk to improve financial outcomes for individuals, organizations, and the public." [**36**] Students may find their Be an Actuary website[16] helpful in learning about a career as an actuary.

National Council of Teachers of Mathematics. The National Council of Teachers of Mathematics[17] (NCTM) is a professional organization for teachers of mathematics whose mission is to "advocate for high-quality mathematics teaching and learning for each and every student." [**27**] They publish books and journals, including the *Journal for Research in Mathematics Education*, that are specifically focused on teaching mathematics at the PK–12 level.

Society for Mathematical Biology. The Society for Mathematical Biology[18] has an annual meeting, often in conjunction with international partners, as well as a journal, *The Bulletin of Mathematical Biology*, and awards. The society is interdisciplinary, serving professionals from multiple departments in both biological sciences and mathematical sciences and industry to academia [**34**]. They also offer Subgroups for deeper specialization, such as Mathematical Oncology and Mathematical Epidemiology, which is the advantage to bringing together a more specialist community.

5.2.3 Specialized organizations supporting marginalized mathematicians.

National Association of Mathematicians. The National Association of Mathematicians[19] (NAM) is a nonprofit organization that seeks to promote excellence in the mathematical sciences for underrepresented American minorities in general and African-Americans in particular [**40**]. Throughout the year, NAM has a number of annual events that support its community.

NAM's Undergraduate MATHFest is a three-day meeting, typically Friday through Sunday in the Fall, which rotates around the country, based on NAM's regional structure. It is held annually to encourage students to pursue advanced degrees in mathematics and mathematics education. The conference is geared towards undergraduates from Historically Black Colleges and Universities (HBCUs), although all are welcome to attend.

Association for Women in Mathematics. The Association for Women in Mathematics[20] (AWM) is a society that promotes the advancement of women in mathematics research [**4**]. It also sponsors a number of outreach and award programs.

---

[15] https://www.soa.org/
[16] https://www.beanactuary.org/
[17] https://www.nctm.org/
[18] https://www.smb.org/
[19] https://www.nam-math.org/
[20] https://awm-math.org/

5.2. Professional Societies

Figure 5.2. Participants at NAM MATHfest. Photo courtesy of Dr. Zerotti Woods.

Many AWM student chapters are quite active on campuses across the nation and can provide an excellent community of support for women who may be in the gender minority in some mathematical sciences programs. Membership is open to all.

Spectra. Spectra[21] is an organization that helps provide "recognition and community for Gender and Sexual Minority mathematicians." [**37**] Its resources include an Outlist—a list of mathematicians that are "out" and identify as a member of the LGBTQ+ community—as well as an ally list. This group often holds receptions at national meetings, and there is a mailing list you can join for more information. They also recently released a guide to LGBTQ+ conferences:[22] "Spaces For All: The Rise of LGBTQ+ Mathematics Conferences."

Society for the Advancement of Chicanos/Hispanics and Native Americans in Science. The Society for the Advancement of Chicanos/Hispanics and Native Americans in Science[23] (SACNAS) is "an inclusive organization dedicated to fostering the success of Chicanos/Hispanics and Native Americans, from college students to professionals, in attaining advanced degrees, careers, and positions of leadership in STEM." [**35**]

SACNAS offers student institution-based chapters, online events and resources, regional meetings, and a national meeting each fall. The application deadline to present a poster at the SACNAS' National Diversity in STEM Conference is many months before the conference itself, and the poster selection process is highly competitive. However, all applicants get feedback from professionals and participants

---

[21] http://lgbtmath.org/
[22] https://www.ams.org/journals/notices/202106/rnoti-p998.pdf
[23] https://www.sacnas.org/

in the national poster session are eligible for awards. Even if you are not presenting, the conference experience is designed to promote success for undergraduate students, including a one-on-one mentoring program opportunity.

5.3 Conferences and Events for Communities of Mathematicians

Attending conferences is a great way to meet people beyond your institution in the mathematics community. Many professional societies host conferences, as mentioned above. There are even more opportunities to gather at conferences and events that are specifically suited to subgroups of mathematicians or certain geographical areas.

Online Undergraduate Resource Fair for the Advancement and Alliance of Marginalized Mathematicians. The Online Undergraduate Resource Fair for the Advancement and Alliance of Marginalized Mathematicians (OURFA^{2}M^2) was founded and is entirely run by undergraduate mathematicians from groups that have been historically marginalized, minoritized, underrepresented, and underserved in the math community. Their main aim is to share information and offer resources to fellow students who are less likely to be plugged in to the math community by virtue of their background. OURFA^{2}M^2 offers a free, virtual conference (typically in November) that gives participants a wealth of useful information about REUs, grad school, and more. To learn more, visit their website.[24]

The Nebraska Conference for Undergraduate Women in Mathematics. The Nebraska Conference for Undergraduate Women in Mathematics[25] (NCUWM) is a fantastic conference that not only provides opportunities to present your research and meet undergraduate math majors from other schools, but it also provides a space for panel discussions on being a woman in STEM.

Field of Dreams. The Math Alliance has a conference called Field of Dreams[26] as well as regional conferences. According to the Field of Dreams website, "The Field of Dreams Conference introduces potential graduate students to graduate programs in the mathematical and statistical sciences at Math Alliance schools as well as professional opportunities in these fields. Scholars spend time with faculty mentors from the Math Alliance schools, get advice on graduate school applications, and attend seminars on graduate school preparation and expectations as well as career seminars."

Mathematical Sciences Institutes Events. The math institutes are a group of National Science Foundation (NSF) Mathematics Centers which stimulate research and training opportunities for the mathematical community. A full updated list can be found on the NSF math institutes website.[27] Most of the opportunities are geared to graduate students, postdocs, or professors; however nearly all of the institutes offer some kind of undergraduate summer research program and/or research conference. In addition, some of the institutes offer live-streaming or

---

[24]https://sites.google.com/view/ourfa2m2
[25]https://math.unl.edu/ncuwm
[26]https://mathalliance.org/field-of-dreams-conference/
[27]https://mathinstitutes.org/institutes/

recorded public lectures, which may be a great way to learn more about a new area of interest.

The Mathematical Sciences Institutes also collaborate to hold the Infinite Possibilities Conference,[28] the Blackwell Tapia Conference,[29] and the Conference for African-American Researchers in the Mathematical Sciences[30] (CAARMS).

Fields Institute Events. The Fields Institute has outreach events, such as LGBTQ+ Math Day[31] and the Queer and Trans Mathematicians in Combinatorics Conference.[32] It used to be that many of these required travel funding, but several are opening up now for free virtual attendance.

Regional events. Independently, there are many regional conferences that are specifically devoted to undergraduate research. These conferences are great places to present your work and meet undergraduate math majors from nearby schools. They often involve social events, such as puzzle hunts, cryptography challenges, or integration bees, to make the event more fun. Frequently, these conferences may be coupled with regional MAA meetings, but sometimes they run independently. Faculty at your home institution can advise you on local options.

5.4 Support from Your School

Many schools and/or departments themselves can support undergraduates in gathering to discuss mathematics, especially students doing research. This may take the form of scholarships, research grants, or travel grants awarded by your department, your dean, or the campus student government. Some of these funding opportunities have application deadlines, so inquire early and then mark it on your calendar.

Many institutions also hold an undergraduate research day where you can present a talk or a poster. Take advantage of this opportunity if you want to present your research at a regional or national conference. The audience may be a little different, but it will help you practice your take-home points and your communication skills (see Chapter 6 for a more in-depth discussion of technical skills).

Finally, think about your communities of practice in your institution. There may be math clubs, a Pi Mu Epsilon or Kappa Mu Epsilon chapter, or local student chapters of some of the professional societies mentioned above (e.g., SACNAS, AWM). All of these organizations can help support you as a person to reach your mathematical goals, and they may also facilitate outreach events that give back to your broader local community.

5.5 Tips on Conferences and Networking

Attending conferences and workshops can play an important role in your development as a mathematician. While attending a conference, you will learn new things, share your research with a broad mathematical audience, and network with other mathematicians and scientists. At the same time, the prospect of attending

---

[28]https://mathinstitutes.org/diversity/infinite-possibilities
[29]https://mathinstitutes.org/diversity/blackwell-tapia-conference
[30]https://caarms.princeton.edu/
[31]http://www.fields.utoronto.ca/activities/21-22/LGBTQplus
[32]http://www.fields.utoronto.ca/activities/20-21/QTMC

a conference can be daunting. There are logistical challenges (How will I get to the conference? Where will I stay? What will I eat? How much will it cost? How will I make up for missed classes? How do I tell them I want to give a talk?), scientific challenges (Is it really five straight days of math? What do you mean I'm supposed to give a talk? What should I do while I'm not giving my talk? What if I don't understand what other people are saying?), and social challenges (What if I don't know anyone? I don't feel comfortable introducing myself to people. I don't want to eat dinner by myself every day.) We will do our best to address these questions, but first it might help to learn about some different types of conferences and why you might consider attending them.

First, let's talk about the difference between conferences and workshops. Typically, conferences are meetings where lots of scientists come together to give talks, share ideas, meet new people, and see old friends. Some conferences are attended by people ranging from high school students to retired faculty from all over the world. Others are more specialized, focusing on a specific geographic area (for example, the Pacific Northwest), mathematical area (for example, knot theory, wave mechanics, or Bayesian inference), or demographic group (undergraduate researchers, women in mathematics, or Latinx mathematicians). Workshops are similar in their goal to bring people together, but they tend to be more focused on a small number of experts sharing their knowledge with a (relatively) large number of non-experts. For example, there might be a workshop for women in pure mathematics graduate programs who want to learn more about data science. Both conferences and workshops play an important role in building mathematical communities and disseminating knowledge, but the means of accomplishing these goals is slightly different.

How do I get there? Unless the conference (or workshop) you are attending happens to be hosted at your home institution, you will have to make a plan for attending it, and that will likely cost money. But don't despair! There are typically ways for you to get your travel expenses reimbursed.

As you are thinking about traveling to a conference, put together a list of your expected expenses. Will you need to fly? Can you drive? Can you carpool with other people from your university?

Once you arrive, where will you sleep? Will you need to get a hotel room, or will the conference provide accommodations in campus housing? If you're going with other students from your home institution, is there any way to share a room with someone to cut down on costs?

Similarly, what will you eat? Will some or all meals be provided by the conference organizers, or will you need to take care of some meals on your own? Many schools have rules about reimbursing per diem food costs while traveling, which you should investigate and add to your list of reimbursable expenses.

If you are going to a big city, take a look at the list of hotels recommended by the conference. They may be a bit more expensive, but they are also less likely to be twenty miles away from the conference center in a neighborhood that's two hours away from the conference center by public transportation. If you need to fly, how will you get to the airport? Will you need to pay for parking or a shuttle bus? Once you arrive at your destination, how will you get from the airport to your hotel? Will you need to take a cab? Is there a public transportation option? Is it safe? This information can typically be found on the conference website.

Once you have an estimate for how much the trip will cost, talk to your advisor about funding opportunities from your home institution. Many schools have ways

5.5. Tips on Conferences and Networking

for students to apply for funding to attend a conference, which may or may not be contingent on you presenting research. If you are presenting research that was done as part of an REU, ask your mentors if the REU can provide you with some travel funds.

In addition, the American Mathematical Society (AMS) and Mathematical Association of America (MAA), along with other professional societies, such as the Association for Women in Mathematics (AWM) have travel grants to support students attending larger conferences, such as the Joint Mathematics Meetings. Each of these funding sources may provide a few hundred dollars, but together they can be enough to cover your expenses. Attending national meetings tends to be more expensive because you are more likely to have to fly to get there and you will need more nights in a hotel. Besides, incidental expenses such as food and ground transportation are simply more expensive in a big city.

I made it to the conference. Now what do I do? This is a very common question. Attending your first conference can be a bit overwhelming. Maybe you're giving a talk or a poster presentation, which will account for an hour—or maybe just ten minutes—of your time at the conference. What should you do for the rest of the time? Here are a few tips to help you navigate your first conference experience.

How will I know what's going on at the conference? Most conferences have published programs (printed on paper or published digitally) with a schedule of talks and events. These resources are typically available before the conference even starts. For large conferences, such as the Joint Mathematics Meetings, these programs can be a bit daunting, literally spanning hundreds of pages. In most cases, the programs are more manageable.

Which talks should I attend? When you look at a conference program, you may find that there are five, twenty, or even fifty different talks scheduled at the same time. How do you decide which one to attend? This is a difficult question. In a lot of cases, the name of the session where the talk is occurring can help you decide whether it is worth attending. Sessions that are specifically focused on undergraduate research can be promising because they should be approachable for an undergraduate audience. A session called "Invited Paper Session on Motivic Cohomology of Perverse Sheaves," probably isn't the best place to look unless you already know about motives, cohomology, and perverse sheaves (or know that you want to know more about them). In addition, most conferences have keynote talks, which are typically scheduled as the first talk in the morning, right before lunch, right after lunch, or at the end of the day. These talks tend to draw good speakers who are tasked with presenting the big ideas behind their research. This gives you an opportunity to learn something new about a range of mathematical disciplines.

Deciding which talks to attend also depends a bit on your interests. Are you going to be applying to graduate programs sometime soon? Is there someone from a school you are interested in attending, perhaps who works in a field you're interested in studying, who will be giving a talk? You should go to that talk! Are there grad students from that school giving talks? You should go to their talks. Are you looking for a job in industry? Are there people from companies that you would like to work for who are giving talks? You should go to their talks!

I just went to a talk and got lost after 5 minutes. Now I feel dumb. Don't feel dumb! You aren't dumb. You're at a math conference! How many people get to say that?

Honestly, this happens to everyone. Not only does it happen to everyone, it happens to everyone most of the time. It is very rare to understand every single detail of any talk you attend, even if you are an expert in the field. Instead, let's reframe your expectations about attending talks. If you take something away from the talk, then attending the talk was worth your time. Maybe you only understood 5 minutes out of a 60-minute talk. But did you learn something new? Did it help solidify your understanding of something you're learning about in one of your classes? Maybe you didn't understand that equation with 16 variables (why are some of them Greek and others are Hebrew?) and maybe you got distracted because the speaker repeatedly talking about the 'limsup' made you think that a bowl of soup sounds pretty good for lunch. But did the slide with a nice picture or graphic help you get an overall impression of the problem the speaker was studying? Small takeaways are victories at conference talks! Maybe you didn't understand that talk about bootstrap percolation today, but after you attend six or seven more related talks and take more classes in statistics, you will start to understand more. Learning math is a marathon, not a sprint.

And if you were completely lost and feel that Jon Snow knows more than you do about this particular area of math,[33] that's ok, too. Maybe it was a bad talk. Learn from that. What were some things the speaker did that were unhelpful to you as an audience member? Work on avoiding those things when you give your talk. Maybe it was a really good talk, and it was just over your head. That's ok, too. What were some good things that the speaker did? Work on getting better at those things in your talk.

It's 10 a.m. on Friday. This conference runs until noon on Sunday, and I'm already exhausted. Now what do I do? That's normal, especially at your first conference. Conferences are exhausting. Your brain is working hard to learn new things, you're trying to navigate an entirely new environment, and it can feel like you need to be "on" at all times. It's draining. You don't need to be doing math all the time. The first time I (SRK) attended the Joint Mathematics Meetings, a friend and I spent an hour riding elevators as high as we could go in the various conference hotels, trying to find a nice view of the city because we were too tired to attend any more math talks. I still graduated from college and got into grad school. Besides, one of the most important aspects of a conference is making social connections. You can't talk with other people during talks (this is generally considered to be rude), but it's ok to skip some talks in favor of making connections with new or old friends. Take some time to explore a new city, or a new campus. If there's an exhibition hall, take some time to stroll around and see what other people are doing. You can pick up some SWAG, see some cool art, or meet with companies that may want offer you a job. You never know who you'll meet or what connections you'll make! And if you need a nap, go take a nap.

Meeting people seems like a good idea in theory, but it's harder for me to do it in practice. Meeting new people at conferences can be intimidating, especially if they are people whose names are recognized, either for proving big mathematical results or writing the textbook you're using in one of your classes. Here are a few tips for meeting new people at conferences.

---

[33] And canonically, he knows nothing.

Let's start with a more intimidating scenario: maybe you are attending a conference and your advisor has tasked you with going to meet some famous mathematician. This can be very scary, but also very beneficial, with the potential to lead to future collaborations, jobs, or grad school opportunities. When approaching a mathematician you don't know, you might start by saying, "I saw the talk you gave about canonical snarks this morning. It was really interesting! I especially liked how you explained some of the more complex definitions with pictures. Is the fact that your field is visual something that drew you to it in the first place?" Or if the mathematician has not yet given their talk, you could say, "I'm really looking forward to your talk tomorrow! I've always wanted to know more about picturesque varieties. Are you going to share any open problems during the talk?"

Maybe you don't feel comfortable walking up to a stranger and introducing yourself. If you are attending the conference with one of your professors, ask them to make the initial introduction, especially if the person you are supposed to meet is a friend of theirs.

We know that taking the leap to join a community or attend an event with a bunch of strangers can be scary, but finding people you can connect with in the math community can be tremendously helpful for you as you progress through your education and career.

5.6 Mathcrostic

Contributed by Doug Ensley

In this type of puzzle, you solve the clues (A–L) in the blanks provided, then copy each letter into the correspondingly numbered blank in the quotation. Understanding the words being formed in the quotation provides additional hints in the answers to the clues, many of which are spread throughout the contents of this chapter.

Clues

(A) Cohosts of the *Mathematically Uncensored* podcast (last names in alphabetical order)

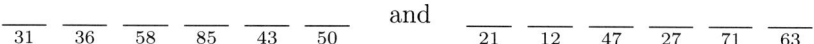

(B) Platform that features contributions of Latinx and Hispanic scholars in the mathematical sciences

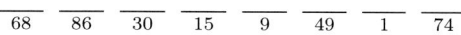

(C) Mathematical bestseller, _____ *Figures* (2016)

78   Chapter 5. Networks and Communities of Support

(D) Mathematical biography, A _____ Mind (1998)

$\overline{84}\ \overline{32}\ \overline{2}\ \overline{81}\ \overline{24}\ \overline{72}\ \overline{52}\ \overline{45}\ \overline{76}$

(E) Inspirations published by AMS and MAA, Living _____: Stories of Resilience Along the Mathematical Journey (2019)

$\overline{61}\ \overline{22}\ \overline{51}\ \overline{89}\ \overline{67}$

(F) Not leaving anyone out

$\overline{69}\ \overline{19}\ \overline{42}\ \overline{79}\ \overline{34}\ \overline{60}\ \overline{23}\ \overline{56}\ \overline{5}$

(G) Appropriately named org promoting "Mathematics for ALL"

$\overline{3}\ \overline{39}\ \overline{20}\ \overline{66}\ \overline{46}$

(H) Louise _____, founding member of AWM, whose award for contributions to mathematics education bears her name

$\overline{54}\ \overline{7}\ \overline{59}$

(I) Name of annual gatherings for both MAA and NAM

$\overline{35}\ \overline{80}\ \overline{8}\ \overline{17}\ \overline{77}\ \overline{55}\ \overline{41}\ \overline{65}$

(J) Popular magazine published by ASA

$\overline{38}\ \overline{33}\ \overline{18}\ \overline{37}\ \overline{10}\ \overline{75}$

(K) Regional gatherings of MAA or SIAM members, where there are often activities specifically for students

$\overline{11}\ \overline{16}\ \overline{78}\ \overline{53}\ \overline{25}\ \overline{44}\ \overline{90}\ \ \overline{6}\ \overline{48}\ \overline{83}\ \overline{14}\ \overline{62}\ \overline{29}\ \overline{87}\ \overline{13}$

(L) Arial or Helvetica, for example

$\overline{70}\ \overline{28}\ \overline{40}\ \overline{73}$

5.6. Mathcrostic

Quotation

$\overline{}_{1}\overline{}_{2}\overline{}_{3}\overline{}_{4}\overline{}_{5}\overline{}_{6}\overline{}_{7}\overline{}_{8}\overline{}_{9}\overline{}_{10}\overline{}_{11}\ \overline{}_{12}\overline{}_{13}\ \overline{}_{14}\overline{}_{15}\overline{}_{16}$

$\overline{}_{17}\overline{}_{18}\overline{}_{19}\overline{}_{20}\overline{}_{21}\overline{}_{22}\overline{}_{23}\overline{}_{24}\overline{}_{25}\overline{}_{26}\overline{}_{27}\ \overline{}_{28}\overline{}_{29}\ \overline{}_{30}\overline{}_{31}\overline{}_{32}$

$\overline{}_{33}\overline{}_{34}\overline{}_{35}\overline{}_{36}\overline{}_{37}\ \overline{}_{38}\overline{}_{39}\overline{}_{40}\overline{}_{41}\overline{}_{42}\overline{}_{43}\overline{}_{44}\overline{}_{45}\overline{}_{46}\overline{}_{47}\overline{}_{48}\overline{}_{49}\overline{}_{50}$

$\overline{}_{51}\overline{}_{52}\ \overline{}_{53}\overline{}_{54}\overline{}_{55}\ \overline{}_{56}\overline{}_{57}\overline{}_{58}\overline{}_{59}\ \overline{}_{60}\overline{}_{61}\overline{}_{62}\overline{}_{63}\overline{}_{64}\overline{}_{65}$

$\overline{}_{66}\overline{}_{67}\ \overline{}_{68}\overline{}_{69}\overline{}_{70}\overline{}_{71}\ \overline{}_{72}\overline{}_{73}\overline{}_{74}\overline{}_{75}\overline{}_{76}\overline{}_{77}$

$\overline{}_{78}\overline{}_{79}\overline{}_{80}\overline{}_{81}\overline{}_{82}\overline{}_{83}\ \overline{}_{84}\overline{}_{85}\overline{}_{86}\overline{}_{87}\overline{}_{88}\overline{}_{89}\overline{}_{90}$

6
Technical Skills

There are all sorts of skills that are useful, directly and indirectly, for having a career in the mathematical sciences. These include being able to use core mathematical content, effectively support claims using data, and form logical arguments. They also include skills that may be self-taught or taught "on the side," such as the ability to give a good technical talk, independently read a research paper, write code to carry out experiments, or typeset mathematics in LaTeX. Many math majors work as tutors and pick up skills that prepare them for teaching and mentorship as well.

In this chapter, we will discuss many of these skills, their importance, and how to acquire them. But first, let's talk about one of life's most crucial skills: how to be someone who people actually want to work with.

6.1 Collaboration Skills

Wait—I thought this was math! Why do we need to have collaboration skills?! All the great mathematicians completed their best work solo, so is it really necessary to be able to work with other people? I (CDE) remember dreaming about a future in a cubicle. And even when my dream became teaching, I thought the job would basically be about how good my math explanations were. Well, it turns out that in practically all jobs, there is an element of teamwork. Most math research these days is done by teams of collaborators. Teachers and professors are part of a faculty, which needs to achieve consensus about policies and curriculum. And in teaching, it has a huge impact on student learning when teachers can develop a relationship of trust with their students. In industry, you have to be part of a team, even if there are aspects of your job that you can do independently.

How does one learn the technical skill of playing well with others? First, don't assume that everyone you're going to be working with in your future career is going to be like you. This is one reason why it's a great idea when you're an undergraduate to take courses that explore humanity, from departments like sociology, psychology, or women's and gender studies. You might take courses that connect how people think to the way the world is constructed, such as a language course or a course in economics, history, or politics. You can also enhance your own personal learning outside the classroom. Many colleges and universities have programs to get involved in supporting diversity efforts, exploring anti-racism, or contributing to projects that address local community-identified needs.

These co-curricular opportunities are also a great way to develop group collaboration skills. Hopefully, some of your classes incorporate things like group work and learning reflections. Avoid the temptation to think of these aspects of class as distractions from your mathematical learning. They are enhancements that are meant to set you up for later success.

As professors, we often hear from students about issues with group dynamics when they need to complete a group project. Part of the problem is that we (professors) may assume you already know how to function in groups. You may also incorrectly assume others in your group have the same approach to group work that you do. Many group work issues can be avoided or become less stressful to recover from if you set your group's structure, roles, and communication expectations at the very beginning. To avoid disappointment later in the process, below are a few suggestions for setting up group projects, particularly those that last multiple weeks. Notice that these tips will require you (or someone else on your team) to step up as an informal group coordinator. Although this role can feel awkward, it's a great role to practice. For instance, in industry, senior project contributors spend a good portion of their time organizing individual responsibilities across the team and reaching consensus on a course of action with different stakeholders.

So, at your first group meeting (typically in class), don't start working on the assignment right away. Instead,

- Start by introducing yourselves to each other.

- Agree on how you will communicate outside of class (text, group chat, email, etc.). Exchange contact information. If someone is hesitant to share personal contact info, respect their boundaries and use school-provided emails/messaging systems.

- Establish norms for communication, such as how often will you check in with each other.

- Read the entire assignment to understand the requirements.

- If there are options for the topic/format of the project, brainstorm. Be respectful of everyone's ideas. The first type of problem that occurs in group projects is when someone feels their ideas are ignored or rejected without real consideration.

- Claim roles and assigned tasks (taking into account preferences and skills). Consider having a backup person assigned for any key roles. Understand also that different people approach problems in different ways—some process their ideas out loud and will generate a lot of ideas that can be evaluated later; others like to take time on their own to think about a problem before sharing their ideas with a group; others are good at working with an existing idea to tinker with and improve it; and others still may need to ask a lot of questions first before they feel comfortable starting in on a problem. Try to understand your style along with your teammates' styles. This can help not only with assigning roles but also in smoothing frustrations that can arise from differing interpersonal dynamics as you start working together.

6.2. How to Read a Math Textbook

- Have everyone agree that it's also the role of each group member to review the final product to make sure they understand everything. Each group member should be prepared to answer questions about any part of the project, if asked.

- Make sure that at least one person will act as the final copy editor. Final editors work to make sure that everything is coherent and consistent (do you use the same formatting and terminology?), and they will contact anyone who needs to make edits to their work. They will submit the project to the professor when it's complete. This is an important role.

- Sketch out a timeline—when should each person have a draft of their portion? Is anyone's task going to be held up by another group member not completing theirs on time?

- Before leaving, decide on the next meeting time. Also, record and share a list of tasks people need to do before that next meeting.

Some collaborative assignments can be completed asynchronously, perhaps using group chat and Google docs or Overleaf. Some assignments require active time brainstorming together. Hopefully, by laying a good foundation for your collaborative work, you'll have a better idea of how often you'll need to meet outside of class.

One final tip to be a successful collaborator: Be excellent to each other.

6.2 How to Read a Math Textbook

At some point, one of your professors will tell you to go and read your math textbook. This is easier said than done. Reading a math textbook is different from reading other books or textbooks. The material is dense and difficult to navigate, and it will often require several readings before you can make sense of it.

In the past, you may have "read" a math textbook by flipping through to find a key problem-solving tool in a callout box or an example that strongly resembles a homework problem you were asked to do. This way of using a book can be helpful for doing specific homework problems, especially in introductory math courses, but it is a less useful approach if you're reading for a deeper understanding or if you're studying more advanced material. So, here are some tips for you to consider to get the most out of that math textbook that you spent three months' rent on.

- Don't attempt to read a math textbook when you're mentally exhausted, distracted, or sleepy. Reading math requires deep concentration. Try to do so during a time of day when your brain is functioning well, and minimize distractions.

- Start by carefully reading the introduction of the section or chapter to help you understand the context of what you're about to learn.

- After you get a sense of the big picture from reading the introduction, skim the section of the book you're focused on to learn what the major definitions and results are. If these are too opaque, at least familiarize yourself with the terms and notation you're about to learn as you dig deeper. Later, read through to try to understand more of the details. (See Section 6.3 on "How to read a research paper" for more detailed recommendations on how to approach your second or third readings. Much of the same advice applies to textbooks, too.)

- Examples are your best friends. If the textbook provides an example problem and solution, read through it carefully to try and understand how it works. Once you more-or-less feel like you understand it, write down the problem on a separate sheet of paper and see if you can recreate the solution or portions of the solution on your own. When you get stuck, you can always look at the worked-out solution for a hint to get you started again.

- Keep a running list of questions to ask your professor, tutor, or TA. For instance, "What does X definition mean?" or "What is another example of Y?" or "I don't understand step Z in the solution. Can you explain it to me in a different way than the book explains it?"

In general, we encourage you to think about reading the textbook as a tool that will help you get the most out of your classes. If you know which topics will be covered on which days, then you can try to pre-read the sections that will be covered during each class. By coming to class prepared with some (albeit vague) idea of the material you're about to learn, your brain will have additional context for understanding it and you'll get more out of class. Then, after class, you can read the same section again. You might be surprised at how much easier it is to understand than the first time you read it!

6.3 How to Read a Math Research Paper

Historically speaking, most of the math you learn through the beginning of your undergraduate career could be classified as "old news." The study of algebra and geometry, which are the pillars of high school mathematics, go back hundreds or even thousands of years. Many of the ideas that became modern calculus were developed in the mid-1600s, and the foundations of what we today call linear algebra were laid in the 1700s.

At some point, your journey through the history of math will take a quantum leap forward, taking you from the time before electricity to the cutting edge of mathematical knowledge in just a few short years. In doing so, your resources will also shift from lectures and textbooks to math research papers that may have been written after you were born... or even just last year!

Reading new math papers is exciting, but it is also very difficult. The authors are attempting to convey completely new ideas. For many historical concepts, scientists and educators have found good ways to explain what were once complicated ideas, but for results that are hot off the press, the "right" way of framing a problem or solution may still need to be discovered. This means reading a math paper is different than reading a textbook or other pieces of literature (each of which can be challenging in its own right).

When reading a math paper, it is important to start by asking yourself, "Why am I reading this paper?" Do you just need to get the main idea? Do you need to understand each line in depth? Do you need to understand everything, or just one specific lemma or theorem? This will help determine a strategy for approaching the paper.

We will start with the assumption that you really need to read the entire paper. Pieces of the advice given here will apply if you only need to get bits and pieces of information from the paper. The focus here is on theoretical math papers, which tend to follow a pattern of "state some definitions and lemmas, state

6.3. How to Read a Math Research Paper

some theorems, prove the theorems." This certainly is not the norm in all fields (particularly machine learning or data science), but the general approach outlined below is still applicable to reading a wide range of scientific papers.

One of the key ways that reading a math paper is different than reading literature is that you should not expect to read the paper from start to finish. Instead, view reading the paper as an ongoing, nonlinear process of increasing your depth of understanding. To do this, you should expect to read the paper at least three times.

The first reading: The first reading of the paper should be focused on the big ideas. Read the abstract and the introduction carefully. What are the authors attempting to do in this paper? What is the main result (the one that would be on the marquee in big letters for the whole world to see)? Why is it interesting? You don't need to read every word. Skip over the proofs, the lemmas, and the corollaries. What's the big idea? Are there other papers that you need to read so that you can understand the context of this paper? Add them to your reading list.

The second reading: Once we understand what the main result *says*, we need to understand what it *means*. So, we need some examples. If the authors of the paper have been kind enough to give you examples, work through them and make sure you understand them. If they haven't given you examples, or if you have already worked through theirs, make up your own.

Start with the simplest example you can think of: If the result is about groups, why does it hold for the group with one element? If it is about graphs, why does it hold when the graph has one vertex or one edge? If it is about $n \times n$ matrices, what does it say when $n = 1$ (where we are talking about numbers) or when $n = 2$ (where we are talking about matrices we know from linear algebra)? If it is about continuous functions, why does it hold when $f(x) = 1$ or $f(x) = x$? Start simple, and then add complexity.

There two main reasons for doing this: First, you really need to see what the theorem is doing by getting your hands dirty. Math isn't a spectator sport, and this isn't a thought experiment. Get a pencil and paper, and really do the math. Draw pictures. Ask questions. Be curious. This is how you learn. Second, by working through examples you will be forced to wrestle with every new idea and definition presented in the paper. This is your chance to make sure you really understand what's going on.

After the second reading, you should have a good understanding of what the paper is trying to accomplish; now it's time to really get into the details.

The third reading and beyond: On the third and subsequent reading(s), we want to put the whole paper together. How is the main result proved and what are its consequences?

Depending on the complexity of the result, it may take one reading to understand the overall strategy of the proof before you dive into the details. For example, maybe it requires ten lemmas and results from six other papers that you also need to go read. Words are written sequentially on a piece of paper, but the proof may not be linear. Draw a chart outlining the dependencies of the proof, making a sort of mental map of how all the pieces fit together.

This third stage may take place over days or even weeks, and it can depend a lot on what you need to get out of the paper. Sometimes, you'll only need to understand a small part of a paper, and you can focus on that. Let's say you only need to understand one theorem. Then start with that theorem and work backwards. Can you understand the entire statement of the theorem, or do you need more information, such as the definition of a "canonical snark" or the meaning of the following notation:

$$\mathcal{X}_{(\ell,n)}^{\pi} \nearrow \mathcal{X}_{(\ell,n-1)}^{\pi} \searrow \cdots \nearrow \mathcal{X}_{(\ell,1)}^{\pi} \searrow \widehat{\mathcal{X}}_{\infty}.$$

Bounce your way around the paper until you understand everything you need to know. Meanwhile, try to get comfortable with the discomfort and confusion you will likely experience along the way. This is a normal part of the process of learning cutting-edge mathematics!

6.4 Writing Math: LaTeX

Anyone who has ever attempted to write a paper containing even a nominal amount of mathematics in a standard word processing program understands the struggle. There are so many menus, so much clicking, and so much highlighting. It's so slow. Now why is my cursor still in the exponent? How do I get out of this box? Which one of these letters is a capital psi? Is that a degree symbol or a copyright symbol? (At this point, our hope is that you feel like you are in the middle of a cheesy infomercial.) *There has to be a better way!*

Fortunately, there is! LaTeX, or TeX for short, (pronounced 'LAH-tech' or 'tech') is a software package that was specifically designed to typeset mathematics. There are innumerable books, blogs, and other resources that you can use to learn LaTeX (we will provide some references below), so we will not attempt to explain how to use it here. Instead, we will give a quick overview of why it's worth investing the time to learn how to LaTeX.

First and foremost, LaTeX was engineered to seamlessly typeset mathematics in a text document, as opposed to most standard word processing programs whose primary objective is to typeset words with added features that allow you to add equations if you absolutely must.

Second, LaTeX is a programming language, which means you write code to display complicated mathematical expressions. For example, if you type `$\int e^{r\theta} d\theta$`, the output file will display $\int e^{r\theta} d\theta$. The LaTeX software interprets the first dollar sign as saying, "Ok, it's time to display some math now;" the command '`\int`' displays an integral sign; and the caret '`^`' is used for exponents, just as it is on your calculator. The '`{`' and '`}`' parentheses hold everything that will go in the exponent, and '`\theta`' is the command for typesetting a lowercase theta.[1] The final dollar sign sends the message, "Ok, now we are done displaying math and can continue with ordinary text." Once you become familiar with LaTeX, it takes no time at all to display that integral, whereas it would take significantly longer to dig through menus to find the necessary symbols in a word processing program.

Third, LaTeX syntax for mathematical equations has also been adopted in Markdown (a very commonly used lightweight markup language used by blogs, online forums, wikis, readme files, and tech documentation). This means that its usage extends beyond math and physics papers!

Finally, LaTeX is more universal than other file formats. Have you ever tried to share a word processing document across different operating systems or computers?

---

[1] A great example of the logic of LaTeX: for an uppercase theta, you'd use '`\Theta`' instead.

Ack! What a nightmare! Because you are only sharing source code in a LaTeX file (or a .pdf), it is much less susceptible to formatting errors when being shared by multiple users.

For all of these reasons (and more), LaTeX is the standard tool used in professional mathematical writing. (In fact, this book was typeset in LaTeX!) The vast majority of reputable math/statistics/data science/machine learning journals and conference proceedings use LaTeX to typeset their papers. Beyond this, many other tools, such as Jupyter notebooks, have functionality that allows the user to enter LaTeX source code in a markdown cell that can be rendered and displayed nicely on the screen.

Overleaf[2] is an online LaTeX editor that also contains a number of very nice templates and tutorials that you can use for free to get yourself started. If you wish to install LaTeX locally on your computer, instructions can be found at The LaTeX Project.[3]

With all this being said, LaTeX is still a programming language, which means there is some syntax you will need to learn in order to become proficient in it. Learning the syntax of LaTeX can be confusing and frustrating at first, but we encourage you to stick with it, even when your code is producing errors or gibberish and it seems like it would be so much easier to just click some symbols out of a menu. Work through the errors (some tips and tools are below). Soon enough you'll groan in frustration any time you have to typeset mathematics and *cannot* use LaTeX .

LaTeX resources. Here are a few resources to help as you are using LaTeX for the first time. Overleaf has a crash course called "Learn LaTeX in 30 minutes."[4] Another good way to learn LaTeX is to borrow an existing .tex file from someone who already knows how to LaTeX and have them show you how to make changes to it. The open source LaTeX Wikibook is another great resource.[5] If you are going to install LaTeX on your personal computer, we recommend starting in the LaTeX Users Group.[6]

If you don't know how to do something in LaTeX, the best solution is to Google it. There's a very good chance that someone else needed to do the same thing. If you don't know the command for inserting a specific symbol, Detexify can be a real lifesaver.[7] Detexify allows you to hand draw the symbol you want, and it will suggest several options for how to make your symbol appear in TeX.

6.5 Technical Writing Skills

Technical writing and communication skills are a prized asset in almost any career path. Improving your technical writing—whether it is for a term paper in a math class, an expository paper for a capstone project, or for your first ever research paper—is worth your time and effort, as it is likely to serve you well in your future career.

---

[2]https://www.overleaf.com
[3]https://www.latex-project.org
[4]https://www.overleaf.com/learn/latex/Learn_LaTeX_in_30_minutes
[5]https://en.wikibooks.org/wiki/LaTeX
[6]https://tug.org/
[7]https://detexify.kirelabs.org/classify.html

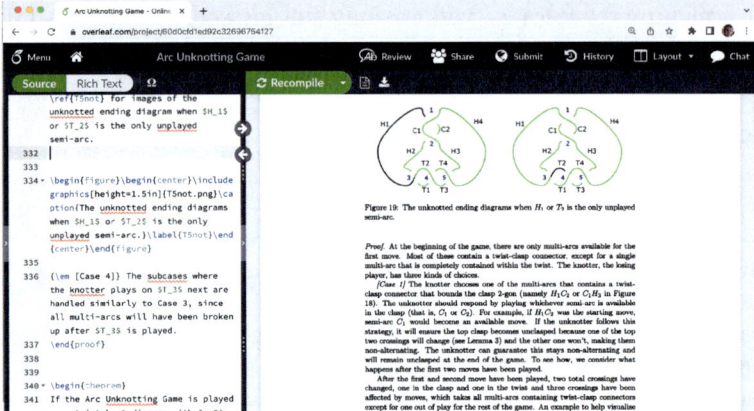

Figure 6.1. A research paper being LaTeXed up on Overleaf.

Here we will outline some broad advice for improving your technical writing. We also recommend the article "Some Guidelines for Good Mathematical Writing"[8] by Francis Su[9] with specific advice for writing in your math classes, which has some aspects that are useful in the short term (such as writing good proofs in your Real Analysis class) and other parts that generalize well to technical writing more broadly.

Draft on paper before you sit down to type. If you are writing your first technical paper, there's a good chance that you're also using LaTeX for the first time. While most of the writing you do in LaTeX is the same as what you would type in your favorite word processing program, having to think about unfamiliar commands while you're trying to figure out the best way to communicate a mathematical idea may involve too much mental multitasking. Consider writing a first draft of the important content, like proofs of theorems, on paper before you TeX up your work.

As a team, agree on notation and terminology in advance (when possible). When you are writing mathematical ideas, you will likely be using a lot of technical terminology and notation. Some of this terminology and notation will be standard, and you can use the same words and symbols as the authors you're citing have used in their papers. On the other hand, you may be introducing new notation and terms in your paper that have no established standard. If this is the case, think carefully about what words to use to describe new mathematical ideas. Your choices for symbols and terminology can either elucidate or obscure meaning. Put yourself in the reader's shoes. Are there ways of communicating ideas that are more natural, appealing to your reader's intuition?

Who is your audience? Knowing your audience will help you decide not only how to structure your paper, but also which details to include and which ones to omit. Despite the fact that your teachers and professors have implored you to show every step of your work throughout your academic career, many small details and calculations are omitted in professional math papers.

As a concrete example, suppose you had to find the value of $\int_0^\pi \sin(x)\ dx$ as part of your paper. If your audience consists of students in an introductory

---

[8]http://bit.ly/guidelines-math-writing-pdf
[9]Author of *Mathematics for Human Flourishing* [39], winner of the 2021 Euler Book Prize

6.5. Technical Writing Skills

calculus class, then you should show every step, explaining that $-\cos(x)$ is an antiderivative of $\sin(x)$, and then showing how to compute the definite integral using the Fundamental Theorem of Calculus. If your audience consists of senior math majors, you might just write, "$\int_0^\pi \sin(x)\,dx = 2$, by the Fundamental Theorem of Calculus." If your audience consists of professional mathematicians, you would just write "$\int_0^\pi \sin(x)\,dx = 2$," trusting that they know how to compute this integral.

This may seem like a silly example, but it illustrates the main point: you need to know who will be reading your work in order to determine how you will communicate. Which aspects of your work require a lot of explanation, and which aspects are standard enough that your audience will understand them without you needing to waste time explaining something they already know? Are there specific styles of writing, pieces of technical jargon, or turns of phrase that are standard in the field that you need to be sure to use? Knowing your audience applies both to your research paper and each presentation you give about your research: you will have to provide more or less detail (or even emphasize different aspects of your process/results) depending on the audience.

Start with scaffolding, and then fill in details. When writing a paper, it can be helpful to work from the inside out. Start by writing out the most technical ideas, such as your definitions and statements of your results. Then fill in the proofs. Then add more details to help the reader understand these technical ideas (like explanations, examples, figures, context, references to other work). Moreover, before you add exposition to help your reader understand your work, make sure you have a clear picture of who your reader is. Again, know your audience.

Independent of your audience, think about yourself when you were first exploring this problem. What helped you understand it better? Use this to think about what examples, figures, and explanations to include to help your reader. A long list of new theorems is nice, but people want to see examples to help their learning. Think of this as the professional version of the answers to the odd-numbered exercises at the back of a trigonometry textbook. Your readers will work through examples on their own to be sure that they understand your results.

Don't be overly rigid or condescending. It is easy to fall into a trap of being extremely formal in your mathematical writing, but this does not always need to be the case. Unless you are writing a legal document, it is generally okay to allow some of your personality to shine through in your writing.

Think of a mathematical paper as an opportunity to tell someone about an exciting project that you just devoted a lot of time and effort to. Structure the paper in the same way that you would explain your work to a friend. Motivation: what makes this work interesting? Background: what information do they need to understand before they can even begin to understand your main results? Examples and intuition: what pictures do you draw when you're thinking about this problem? What are the intuitive ideas that led you to the main result? Tell your reader a story rather than laying out a laundry list of definitions, theorems, and proofs that are devoid of context. This not only makes the paper more fun to write, but also makes for a more enjoyable experience for the people who are reading it.

Similarly, avoid the classic mathematical microaggression of using words and phrases like "clearly," "obviously," or "it is easy to see" when you are writing your results. There are a lot of pieces of the project that may be clear, easy, and obvious *to you* because you've spent a long time exploring these objects. That is not necessarily the case for your reader. There is no need to speak so condescendingly to your reader. After all, they are reading your paper because they care about what you have to say.

Start sentences with words, not symbols or numbers. When writing mathematics, symbols typically act as nouns in a sentence, for example, in saying "Let n be a positive integer," the subject of the sentence is n. Unlike writing prose, however, a symbol should not be the first word of a sentence, and neither should a number. One reason for this is that the period that terminates the previous sentence could be confused with a decimal point or other mathematical symbol, making it difficult for the reader to grasp the meaning of what you have written. There are plenty of ways to restructure a sentence that begins with a math symbol or number so that it will start with a word instead. For example, "Δ is a simplicial complex and therefore..." can be rewritten as "Because Δ is a simplicial complex, we know..."

Be consistent about capitalization of notation/labels. The difference between upper- and lower-case letters in mathematics is very important. For example, we might write, "Let X be a set of real numbers and $x \in X$." Here, the upper-case X represents a set of objects, whereas the lower-case x represents an element of that set. If we go on to write something like $|X|$, we mean the cardinality of set X, whereas $|x|$ is the absolute value of the number x. These distinctions are important! Because of this, it is important to be consistent in the notation you are using in your writing and proofs. Not only can it be confusing to write X when you meant to write x, it can also completely change the mathematical meaning of what you wrote. Correct proofs can become incomprehensible or even incorrect. Civilizations can crumble. Puppies and kittens will be sad.

Cite others' work by name. More often than not, your paper will reference work that other mathematicians have done. Typically, the items of your bibliography will be numbered, for example as:

[1] A. Cayley, A theorem on trees, *Quart. J. Pure Appl. Math.* 23 (1889), 376-378.

When you reference this article in your paper, the citation [1] should not serve as a noun in a sentence. The paper did not do any work. The author(s) did. The citation merely serves as a reference to a paper in the bibliography. This is easier to see in an example.

Good: "Cayley [1] proved that the number of spanning trees in a complete graph on n vertices is n^{n-2}.

Not so good: "In [1], it was shown that the number of spanning trees in a complete graph on n vertices is n^{n-2}.

Even worse: "[1] showed that the number of spanning trees in a complete graph on n vertices is n^{n-2}.

Can you see the difference? In the first example, the author, Cayley, was given credit for the result. In the second example, we are given a reference to a paper, but we still need to turn to the bibliography to determine which paper it is. In contrast, experts who see "Cayley [1] proved..." will immediately know which paper you are referencing. Finally, in the last example, the paper itself is treated as the subject of the sentence, and it seems that the paper deserves the credit for the results it contains as opposed to the author who wrote it.

Finally, for papers with multiple authors, give credit based on the order in which the authors appear in the original work. It is typically the case in theoretical mathematics that authors are listed in alphabetical order. In applied mathematics, this is often the case, but not always; and in statistics and other scientific fields, alphabetical authorship is less common. If there are two or three authors, give credit to all of them, for example as "Provan and Billera [4] proved..." or "Kalai, Nevo, and Novik [14] showed..." When there are more authors, use the name of the first author followed by "et al.", as in "Adams et al. [1] showed..."

Get feedback and be prepared for criticism. Once you have written a draft of your paper, share it with some different people. In addition to your advisor, faculty mentor, and/or collaborators, think about sharing it with some of your peers or other folks from your target audience.

The first time you show your writing to your faculty mentor, be prepared for it to come back looking like a document that has been redacted by the CIA. There will be red ink everywhere and it may seem that the only original words that have survived are (some of) the articles, and perhaps the odd mathematical symbol. This is normal, and a sign the faculty mentor is trying to help your results shine! Communicating mathematics through writing is very difficult, and there are certain bits of written mathematical culture that can only be learned through experience.

Don't let this get you down. It is not a measure of your self worth, nor was it a waste of time to write a lot of things that ultimately got cut from your paper. Many professional writers will produce pages upon pages of text only to have a few paragraphs that end up being used. The act of writing is important. Even if your writing doesn't make the final cut, you likely gained a deeper understanding of your work through the writing process. Getting feedback from a faculty mentor is often the best way to improve your writing.

Read what you have written out loud. When we are editing our own work, it is easy to read what we meant to write instead of what is actually written on the page. By reading what you have written out loud, you are forced to take on the role of the reader. It makes you more likely to catch grammatical errors or find phrases that just don't make sense as you have written them.

6.6 How to Give a Math Talk

Preparing the talk. The traditional standard for giving math talks at conferences used to be "chalk talks," that is, writing down the information you want to convey on a chalkboard, live in front of an audience. That has changed over the years.[10] These days, while you may still see a chalk talk every once in a while, 99% of math talks are given using slides projected from a computer. In fact, at some big conferences like the Joint Mathematics Meetings and MathFest that are held at conference centers and hotels, there is actually no option for writing something during the talk unless you have a tablet or similar technology you can connect to the projector. There is no chalk. There are no blackboards or whiteboards. There is no overhead projector, and you're unlikely to have access to a document camera.

---

[10]When I (AKH) gave my first math talk as an undergraduate, I used hand-written transparencies on an overhead projector. *Gasp!*

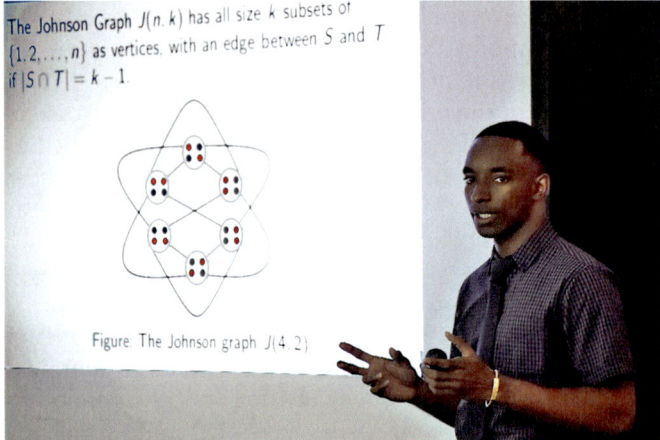

Figure 6.2. A student speaking in the 2018 MSRI-UP final presentations. Photo courtesy of Paul Samuel Ignacio / SLMath.

If you're lucky, there might be a jumbo pad of paper on an easel lurking in the corner of the room, but don't count on it.

So, when you're starting to prepare to give a math talk, the first thing you'll want to do is create your slides. You can use the Beamer package in LaTeX (Overleaf can help you get started with useful talk templates), or you might opt for a more universal platform, like PowerPoint or Keynote. When you sit down to create the scaffolding for your talk and your first slides, here are some things to consider. If you just finished reading the section on writing a research paper, you'll note a lot of similarities!

- Your main goal is to tell a story that facilitates understanding. How do you plan to organize your talk? What is the most logical order for presenting the concepts and results you'd like to share?

- At the beginning of your talk, how are you going to provide your audience with context for understanding the questions you've been thinking about and why they are interesting or important? How are you going to make them care about the topic you're speaking about?

- How are you going to communicate any important definitions or facts your audience needs to understand before they can understand the main results in your talk? Will you provide examples to illustrate different definitions? (Hint: Yes. Yes, you will.)

- Can images, animations, or videos help you communicate certain ideas?

- Which results will you include and which results do you need to leave out, given the time constraints of your talk and the interests of the audience?

- Who do you need to acknowledge at the beginning or end of your talk? Are there any professors, organizations (like the NSF, NIH, CURM), donors, community partners, businesses, or fellow students you should thank? Are there any resources you used that you should mention? Don't forget to thank the conference/session/event organizers and your audience, too!

6.6. How to Give a Math Talk

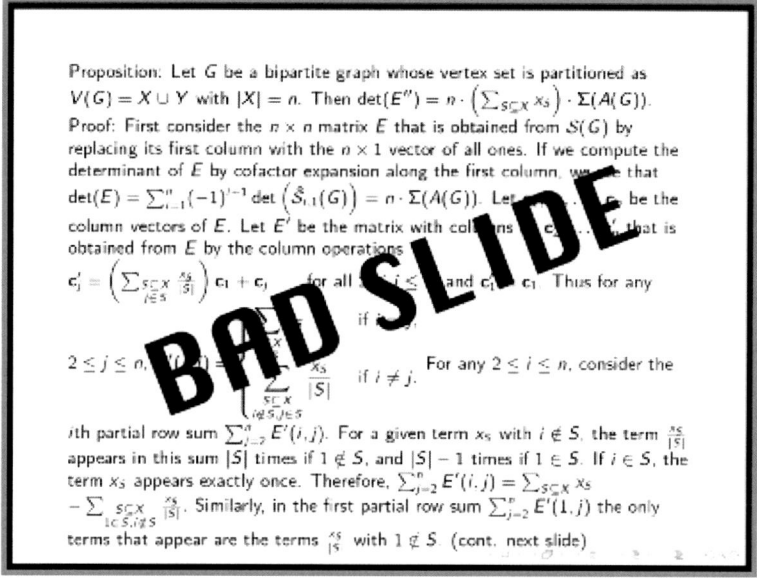

Figure 6.3. An example of a bad slide. Can you find five things that should be avoided in this example?

This list of questions will allow you to think through what content you want to share and how. Here are some additional tips to consider when you're actually creating slides.

Don't put too much information on a slide! For example, if you want to share a proof of a theorem, for Hypatia's sake, don't just copy/paste the proof from a paper onto a slide as in Figure 6.3. Your audience can't listen to you talk and try and read a technical treatise at the same time. Instead, consider giving proof outlines or proof ideas instead of complete proofs. Turn complete sentences into sentence fragments organized by bullet points, and consider revealing each bullet point one at a time on your slide. (Pro tip: \pause is your friend here if using LaTeX/Beamer.) Illustrate proofs or claims you are making with examples to make technical ideas more concrete and understandable.

Plan to have one idea per slide. Use multiple slides to make comparisons. This will make it easier for your audience to follow along and take in the new information you are sharing with them.

When you can use images to illustrate an idea in place of words, do it. There is certainly some truth to the saying "a picture is worth a thousand words." You simply don't have time to belabor a technical point in a 10- or 15-minute talk. Is there an image you can show that will illustrate a definition or result more succinctly?

Use distinguishable colors and patterns. You don't want your audience to miss an opportunity to understand a point you're making because you've used two colors to make a distinction that don't actually look different from each other when they're projected on a screen. Consider using colors with different saturation/hue and different fill-patterns to further distinguish objects so that they'll be separable even for folks who are colorblind (and as a side benefit, even when reprinted in black and

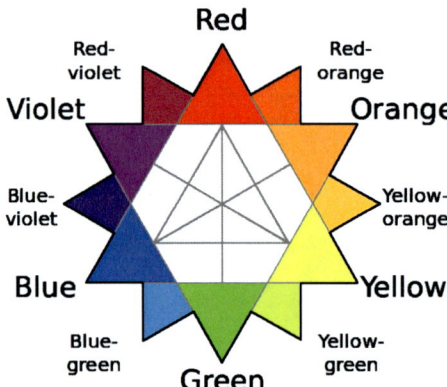

Figure 6.4. A color wheel. Credit by Al2 - Own work, CC BY-SA 3.0, https://commons.wikimedia.org/w/index.php?curid=3049908.

white). Also, consider colors that are opposites on the color wheel (see Figure 6.4) to ensure distinguishability.[11]

Make your fonts easy to read. Will the people sitting in the back of the room be able to read everything you've written on your slide? Yes? Good!

Avoid slides full of complex mathematical notation. If understanding equations is critical to understanding your work, use them and explain them: but don't make your audience have to wade through an entire slide full of complicated equations without some guidance.

I (AKH) once saw a phenomenal applied math talk by an undergraduate researcher on the differential equations that govern how certain water waves move. There were some very complex equations on one slide, but different parts of the equations were underlined in different colors to give a visual cue about how the equations related to one another. The color coding also helped the speaker refer to different parts of the equations to explain their purpose. This talk was great, but rather unique in how well equations were used. More often than not, when I attend applied math talks that rely heavily on complex differential equations, I am overwhelmed by slide after slide of equations that I don't understand. The moral of the story: be thoughtful about how and when you use complex equations!

Preparing yourself. If only your slides could give your talk for you! Wouldn't that be great? For many students, the idea of giving a technical math talk to an audience of peers and professors is daunting. Many of us are terrified of public speaking. Did you know that public speaking is America's biggest phobia?[12]

I (AKH) once had a panic attack in the middle of giving a math talk. It took every coping mechanism I had to avoid fainting mid-talk. My vision was going black.

---

[11] Disclaimer: Try to avoid using colors that are insensitive, like baby-blue and blood-red on a graph about survival rates for babies. Also, avoid confusing color choices, like green to represent something bad and red to represent something good.

[12] http://www.tinyurl.com/mt5jjstz

My heart was pounding out of my chest. My lips were glued to my teeth, and I couldn't breathe or speak normally. That experience taught me that something about my speaking preparation had to change. Since then, I've learned that the best way to make a math talk outstanding and dispel those public speaking jitters is to **practice** and **get feedback**.

To avoid the same nightmare I endured, try these techniques! After you make your slides, give your talk to yourself out loud with as much energy as you plan to muster for the big event itself. Do this at least two or three times. Giving practice talks will help you figure out what you want to say at the beginning of the talk, how you want to make verbal transitions between different ideas, and how you want to explain more complex ideas. You won't get nearly as nervous giving a practice talk to an empty room (although don't be surprised if your stomach is still tied in knots!), and you'll have more brain power to work out some of these verbal kinks. Giving practice talks the same way that you plan to give them to a live audience will give you confidence that you can take with you to a conference.

Next, once you've given yourself a practice talk once or twice, rope small groups of friends into watching you give practice talks. Give practice talks to kind people who you know will smile and nod at you through the talk as well as people who are skilled public speakers themselves who will give you honest, constructive feedback. This will give you more opportunities to rehearse while helping you to improve your talk in ways you might not have thought of yourself. Once you understand your subject really well, for instance, it is easy to take certain knowledge you have for granted. Having people watch you give a practice talk is a great way to test whether or not you've included enough background material for your audience to be able to understand your talk.

When we (AKH and SRK) ran an REU, we required all of our students to give weekly talks on their research progress. Each week, we'd all provide the speakers with suggestions for improvement that could be incorporated the following week. Universally, the talks were quite rough the first couple of weeks. But by the end of the summer, every single participant in the program was a confident, skilled public speaker—even those who came into the program with serious public speaking anxiety.

If you put some time into practicing and getting feedback on your talks, you will be a more confident, more proficient speaker. And remember: public speaking isn't a useless skill. Being able to communicate complex ideas to an audience is an ability that you can use in a variety of contexts throughout your life. So, it's worth the effort you put into it.

6.7 How to Give a Poster Presentation

There are several conferences where it is more typical for students to share the results of their research through presenting a poster rather than a talk. For instance, there is typically a huge poster session at the Joint Mathematics Meetings, co-sponsored by the American Mathematical Society (AMS) and Pi Mu Epsilon (PME), a math honor society for undergraduate students. You'll also find poster sessions at MAA MathFest, at meetings of the Society for Industrial and Applied Mathematics (SIAM), and more.

Figure 6.5. A participant from the MSRI-UP REU shares her research at the 2019 Joint Mathematics Meetings. Photo courtesy of Maria Mercedes Franco.

Preparing your poster. Preparing a poster might seem like a daunting task if you've never done it before, especially if your research mentor asks you to create it in LaTeX. Never fear! If you've had a chance to check out Overleaf to write up your results or prepare a talk, then you'll be happy to know that Overleaf has templates for posters as well. If you don't use Overleaf, you could also ask your advisor or ask around in your department to see if anyone who has made a poster before would be willing to share their source files for you to use as a template.

A good template will include spaces for things like your poster title, your name, and your list of collaborators; your school logo and/or the logo of the organization that funded your work; sections for background material, methods, and results; and a space for acknowledgments. Once you have a good template to work with and you've created scaffolding for yourself, consider these questions to help you figure out what content should go on the poster.

- What are the top one or two most important results you want to share with your audience?

- What is the most important background material you need to introduce so that someone can understand your results and get a sense for the methods used to derive them? Keep in mind that poster session attendees will need to understand your results in just a few minutes. What can you share in this amount of time that will help them get up to speed quickly?

- What visual elements can help you communicate your results and the background for your project?

- Who do you need to thank (e.g., the university where you conducted your research, the grant agency that funded it, your advisor)?

6.7. How to Give a Poster Presentation

When you're preparing the poster itself, keep in mind that people will be standing a few feet away as they engage with it. So, avoid using fonts that are too small, and aim to break up text with relevant images when possible. Also, while you'll want conference goers to get a very basic understanding of the tools that you used to arrive at your results, remember that they won't have time to read through a long formal proof or sift through piles of data. Many of the details related to your work likely won't show up on your poster. For those people who take a particular interest in your research, consider including a QR code that links to a website where they can get more information, for instance, a link to your paper on arXiv,[13] a GitHub repository,[14] or another webpage associated with your work.

Also, keep in mind that your poster should be able to communicate the big ideas on its own. For example, if you leave during the poster session to go get a drink of water or to see a friend's poster, you want someone who walks by your unattended poster to be able to get the big idea from the information you have shared.

After you've prepared a draft of your poster (which you will have done *weeks* before the conference, right?), send it to your advisor or a mentor for feedback. Give them as much time as you can to review it and provide suggestions so that you will have a good amount of time to edit your poster and get it printed before the conference.

While we're on the topic of printing, let's think through some details. We're begging you: don't wait until the night before you fly out to the conference to find out where you will print the poster and if it will cost money. Do this in advance. If you don't have a poster printer you can use for free on your campus, ask your advisor about where to go and how to pay for the printing. Also, find out about how to get a tube to transport your poster to the conference. Do they have poster supplies in the math department office? Will you need to buy a tube at a printing store? How much will it cost and who will pay for it? Please don't wait until the last minute to find out the answers to these questions. A little detective work ahead of time can save you a lot of stress and money later!

Preparing yourself. Now, imagine yourself standing in the poster session having just put up your poster on an easel. The organizers of the session announce they're opening up the room for conference attendees to come talk with you. If someone comes by your poster, do you know what you'll say?

Talking about your work, using your poster as a key aid, can feel like giving a glorified elevator pitch. In fact, it might be helpful to think of giving a poster as somewhere on the presentation spectrum between giving an elevator pitch and giving a talk. If you had 30 seconds to describe the motivation for your work, what would you say? What problem is your research addressing? What got you interested in working on this problem in the first place? Consider how you will be able to connect with someone who is knowledgeable and likes math but who isn't familiar with your particular research area. How will you get them interested in hearing more?

---

[13]ArXiv (pronounced "archive") at https://arxiv.org/ is a place where researchers often share their papers before they've been peer-reviewed and published.

[14]GitHub (https://github.com/) is a platform that enables people to easily share code or software they've developed.

Next, what concepts do you need to define or explain to the conference-goer you're chatting with so they can understand your main result? How can you explain these things less formally than you would describe them if you were giving a longer talk?[15] Is there an analogy or example that you can use to help illustrate your point? How can the poster you've thoughtfully designed help you communicate these ideas? Whatever plan you have for your explanation here, try to keep it to 1–2 minutes.

Finally, if you've done your job well in the first couple of minutes, you will have built a little suspense by the time you talk about your main result and how you arrived at it. It might feel like getting to the punchline of a joke. How can you clearly state your findings, perhaps in a less formal way than the way you've written it in your research paper?

Throughout your poster "pitch," the person you're talking to may ask clarifying or follow up questions, so while your presentation will intentionally omit many of the details of your work, be prepared to talk about these details if you're asked. Make it clear on your poster how the person can connect with you after the conference if they take a greater interest in your research. If you don't include a QR code, make sure to at least include your professional contact information on the poster.

Just as with giving a talk or an elevator pitch, it's important to practice what you'll say to people who are interested in hearing about your work in a poster session. Be sure to run through your spiel several times and time yourself. Practice by talking out loud to yourself in an empty room. Then, practice on your physics major friend. This is doubly important if you are presenting the poster with a partner. Each person should be able to present the poster from start-to-finish on their own, but you should also practice presenting together to know who will be responsible for what and how you will "pass the baton" between speakers. Finally, enjoy the fact that you're well prepared and have fun at the poster session! You'll do great!

6.8 Programming Skills

> "Programming is a skill best acquired by practice and example rather than from books"
>
> – Alan Turing

The ability to use a computer to do mathematical or scientific work is a requirement in most modern STEM fields. Applied mathematicians and statisticians are typically required to take more programming courses than theoretical math students, but more and more algebraists, number theorists, combinatorialists, and topologists use computers to help run experiments and test theorems.

In industrial settings, it would be hard to find a job that did not expect you to have some familiarity with computational tools, which might mean the ability to work in an Excel workbook, query databases in a structured query language, write scripts in R or Jupyter notebooks, or develop machine learning models using Python libraries. Because of this, we generally advise students to take at least

---

[15] For the love of Maryam Mirzakhani, do NOT be tempted to literally read equations.

6.8. Programming Skills

one programming course even if it is not required for their major. If you have the interest and schedule availability, take more than one programming course.

I (JLT) was scared to take computer science as an undergraduate as I thought I was too late to pick up the skill and that I'd be in a class with a bunch of people who already knew how to code. Luckily, my peers convinced me to give it a shot. Because of that computer sicncc course (and the additional ones I took once I learned how fun it was), I got my first post-undergraduate job, got into a PhD program with a computational theory emphasis, was able to take machine learning courses during graduate school, and became prepared for the career I have today. In addition to courses in the computer science department, other exposure to programming can often be gained through courses in applied mathematics, data science, and statistics. If you want to learn to code but aren't able to take any programming courses in a traditional setting, there are a number of free or very cheap online classes (for example, through Coursera[16]) where you can pick up some of the basic skills. Once you understand the basics, it helps to find an independent project where you can apply your skills—it may take three days and fifty StackOverflow articles to get your first program to run, but with time and practice you will gain deeper fluency. If you are looking for problems to start off with, generations of mathematics students have used ProjectEuler as a collection of math puzzles and problems that can be solved by writing short scripts.[17]

As you gain more programming experience, it becomes easier to learn new programming languages. Languages have lifespans and trends, so the key is really about learning to think computationally about problem solving and to learn the general structures that are available to you through programming.

Below is a brief guide to some common programming languages that are used in the mathematical sciences. The emphasis on the languages below is on open-source software (OSS), which is free software that is maintained and developed through mostly volunteer efforts. This means they are still accessible after your academic license for a single course runs out, making the skills you learn more transferable.

Python is a popular object-oriented language, similar to Java or C, but its structure is less syntactically rigid, which has made it a popular first programming language for many students. Because it has a lower barrier to entry, Python has also gained wide adoption in applied settings, data science, and machine learning.

R was developed by and for statisticians. It is an array-based language, meaning that computations are optimized around using structures like vectors and matrices and operations that act on those structures to easily handle and manipulate data. Some packages are available to utilize object-oriented programming when necessary, but this is not its strength. R is well known for its support base for gender-minority programmers (#RLadies).

Matlab/Octave are standard among engineers and mathematicians. Matlab is proprietary software that may or may not be present in your future work. Octave is the open source version that can often stand in when you are required to use Matlab for a class, but development for it is fairly limited. It is an array-based language.

Jupyter[18] is not a language, but rather an open-source project for interactive and collaborative code-sharing across major data science and scientific computing

---

[16]Search for free coding courses at https://www.coursera.org/.
[17]Find fun mathy coding challenges at Project Euler (https://projecteuler.net/).
[18]https://jupyter.org/

languages, like Julia, Python, and R. Jupyter supports "notebooks" that intersperse code, code outputs and visuals, and formatted (markdown) text. Jupyter notebooks are a common way of presenting and sharing analysis, code examples, and data science reports.

SageMath is an OSS computer algebra system as opposed to an independent programming language.[19] SageMath is built in Python and brings together a large set of libraries (or packages) that were created to carry out specialized mathematical tasks. For example, there are libraries for doing calculations in groups, row reducing matrices, walking around graphs, or finding generators for a quotient of a polynomial ring by an ideal.

6.9 Conclusion

We'll end this chapter and our discussion of technical skills by sharing three job postings we found online. The first is for a data scientist, the second is for an entry-level actuary, and the third is for a science writer. In each one we, have **boldfaced** the job requirements that are related to skills discussed in this chapter: collaboration, communication, programming, and writing. Notice how much of the posting is focused on these skills as opposed to the parts that are "just" technical knowledge!

Position 1: Data scientist.

Key job responsibilities:

- Drive design and development of product classification models and other predictive models to improve safety and compliance mechanisms.

- Improve upon existing methodologies by developing new data sources, testing model enhancements, and fine-tuning model parameters.

- **Collaborate with product management, software developers, business intelligence engineers, and business leaders** to define product requirements, provide analytical support, and **communicate feedback**; develop, test and deploy a wide range of statistical, econometric, and machine learning models.

- **Communicate verbally and in writing to business customers with various levels of technical knowledge**, educating them about our solutions, as well as sharing insights and recommendations.

Basic Qualifications:

- Bachelor's degree in Statistics / Applied Mathematics / Operation Research / Engineering / Data Science or other related quantitative fields.

- 2+ years working experience as a Data Scientist or related positions.

- **Proficiency in scripting, querying and/or analytics tools**, such as Python, R, SQL or similar.

- Proficient with various machine learning techniques.

---

[19]https://www.sagemath.org/

6.9. Conclusion

- **Excellent verbal and written communication skills** with the ability to effectively advocate technical solutions to scientists, engineering teams and business audiences.

Position 2: Actuarial analyst.

Education: Applicants must possess a Bachelor's or higher degree in Mathematics, Computer Science, Actuarial Science, Statistics, Economics, or related analytics-oriented discipline. At least 1 Actuarial exam completed (P preferred) and dedication to reaching the Fellowship level.

Experience: No prior work experience required.

Skills/Competencies:

- Intermediate **programming skills**, including VBA, SQL, or related.
- Strong spreadsheet skills in Microsoft Excel.
- **Proven mathematical, analytical, and quantitative skills.**
- Demonstrated ability to **solve problems systematically and creatively**.
- Constant attention to detail.
- Excellent **listening and communication skills, both written and verbal**.
- Ability to work both **independently and as part of team**.
- Effective **time management skills**, including the capability to balance multiple projects with competing deadlines.
- Excellent judgment and decision-making ability.

Position 3: Science writer.

Qualifications:

- At least two years of experience as a journalist covering computer science or a related field, preferably for a major science publication
- **Familiarity with computer science concepts** and research and a demonstrated knowledge of current developments
- Must be comfortable writing news and feature articles under web deadlines and be fluent in the best print editing standards and online conventions
- Must be detail-oriented, organized and an **excellent communicator** demonstrating command of the **highest standards of written English**
- Must be an **excellent communicator and team player who thrives in a collaborative setting**
- Must be **eager to grow** as both a reporter and a writer

Responsibilities:

- The ideal candidate will be an accomplished science journalist eager to **scour journals and engage with experts to learn about the latest discoveries and important trends in computer science**, before presenting it as engaging and accessible prose for a general audience

- **Pitch, report and write** news articles, features, interviews and explainers about theoretical computer science

- Develop a stable of expert sources to help stay informed about the most important and interesting research happening today

- **Collaborate with editors** throughout the editorial and fact-checking process and partner with the design team in the development of artwork, graphics, videos and other associated features

- Occasionally write about **other science and math topics** as needed

- **Attend and report from scientific meetings and conferences**

Part 3

Life After Graduation

7
Careers for Math Majors

> "What are you going to do with a math major?"
> – Someone who doesn't know better

How many folks do you know who have graduated with degrees in math, applied math, data science, or statistics? What are they doing now? Have you ever spoken to an actuary or a science writer? How about an operations research analyst or a lawyer who majored in math? Think about the careers of former math majors you either know well, have met, or have seen give a talk or speak on a panel about their job. Take a look at the BINGO board on the next page and mark a square if you have a relevant connection to someone in that profession.

Did you get BINGO? If not, how could you expand your network to meet folks who studied math and who are now doing different types of jobs? Could you go to a career fair to meet people who work for local companies? Or have you thought about attending a talk in a colloquium or seminar series at your school? You could advocate for the math club at your school to host an alumni panel. You might also consider asking your professors for advice on how to connect with folks in different professions. Sometimes, your professors know about upcoming conferences, seminars, or workshops that may be of interest to you. If you're reading this book with some of your peers, consider having a friendly competition to see who can get BINGO first. You might even find a way to cross a profession off of the board by finding a relevant YouTube video to watch. For instance, if you watch the YouTube video from the PIC Math program on creating more realistic animation for movies,[1] you could cross "Animator" off of the board.[2] What about the careers that aren't on this BINGO board? What jobs would you like to learn more about?

---

[1] https://bit.ly/PICMath-animation

[2] Ok, so Dr. McAdams is technically a Senior Software Engineer, but she works for Walt Disney Animation Studios on visual effects. So, close enough!

Figure 7.1. Math jobs BINGO.

7.1 What Can You Do with a Math Major?

An undergraduate degree in mathematics prepares you for a wide variety of entry points into the workforce. For some career paths, such as teaching or actuarial/data science, majoring in math is a natural choice. For other careers, like those in medicine and art, majoring in math might be a counterintuitive choice. While a major in mathematics might not be the most direct route to achieving your goals, it can give you a solid foundation for your future and open you up to a whole host of exciting career options.

When we reached out to professionals who started with a math major (including lawyers, STEM journalists, data scientists, analysts, composting experts, actuaries, researchers at government labs, software engineers, and business managers), *most said their undergraduate work did not explicitly prepare them for the career they ended up in.* Instead, their mathematics degree helped them develop quantitative and analytical skills, creative problem solving ability, and it ensured (and demonstrated) that they were capable of learning skills and technical content quickly.

In this chapter, we compile information about a wide variety of career options by sharing our own experiences and talking to professionals in many different fields. We'll learn about what drew people to different careers, how their math major helped them prepare for their work, and what advice they have for others who are interested in following a similar path. If you want to learn about even more career options, we recommend Haunsperger and Thompson's *101 Careers in Mathematics* [15] or the *BIG Jobs Guide* [23] by Levy, Laugesen, and Santosa.

While the careers we highlight here don't cover every possible career option for folks with a math degree, you'll get a sense of the breadth of possibilities. One more note before we jump in: this chapter is dedicated to describing career options *other than becoming a professor*. For more on graduate school and academic careers, see Chapter 9.

For some positions you might be interested in (e.g., lawyer, data scientist, researcher), you'll need additional training with at least a master's degree. Others (e.g., analyst, actuary, software engineer) can be accessible directly after completing an undergraduate degree, especially if you take relevant supplemental coursework during your time as an undergrad. So, the chapter on graduate school may be relevant to you even if your goal is not to become a professor or academic researcher.

7.2 Data-Oriented Careers

Consider a music streaming service (like Spotify) which is looking at product usage across the globe. They have terabytes or even petabytes of data that captures many dimensions of user activity and product telemetry (like errors and latency of their product). Data analysts and data scientists (together with data engineers) are responsible for distilling this data in a way that it can be used to make decisions for the product moving forward. The following projects show representative work in different data-oriented roles, though in truth the boundaries between the roles are much fuzzier.

A **data analyst** may compose different data sources and build a "dashboard" report for visualizing product usage across different regions. In this work, they identify an issue, like: most new users in region X are not using the service more than one or two times.

A **data scientist** may dig deeply into this problem, using causal inference and machine learning techniques to find "hidden" relationships in the data. They then isolate and test key likely causes. In our example, a data scientist might learn that recommendations for music in the primary language spoken in region X are particularly poor.

A **data/machine learning scientist** may work to create, train, test, and deploy into production an updated recommendation model. For instance, a data scientist might suggest a new method which improves the recommendation accuracy for music in region X.

Since the mid 2010s, many students of mathematics and other quantitative fields have found employment as data analysts and data scientists. Data scientists and analysts are often pointed to collections of messy data and asked to process it, transform it into useful summaries, extract insights, and build data-based tools or models. The tools and techniques used by data scientists and analysts draw from

Figure 7.2. Stephen Lee works as a data scientist. Photo courtesy of Karla Hernandez. (left) Bethany Lusch works as a data scientist. Photo by Adam Lusch. (center) Ranjani Sundaresan works as a data analyst. Photo courtesy of Arvind Ramesh. (right)

software engineering, statistics, probability, machine learning, and data visualization. Various industries employ data scientists, though the largest employers are tech companies.

Entry-level positions in data-oriented areas often require (1) a degree in Computer Science, Mathematics, Statistics, Economics (or another quantitative field) and (2) experience with a data querying language, like SQL.

For **data analysts**, the degree requirement is usually a Bachelor's, and often communication skills and familiarity with data visualization tools like Tableau or PowerBI are listed as required.

For **data scientists**, the degree requirement is usually a Master's, although some positions are open to those with no graduate degree. Experience with a data scripting language, like Python or R, is required. Problem-solving skills are commonly listed and assessed in interviews.

Given the substantial overlap of skills, it's not uncommon for data analysts to transition to a data science role after a few years of experience. There are also research-oriented Machine Learning positions that require at least a Master's and often a PhD.

We spoke with a few data scientists/analysts who majored in mathematics as undergraduates to better understand their work and how interested math majors can prepare. Interviewee Stephen Lee (SL) worked first as a business analyst in energy markets before transitioning to data science in tech, and he now works in the hospitality sector. Bethany Lusch (BL) works as a data scientist (although her title is "computer scientist") at Argonne National Laboratory. Ranjani Sundaresan (RS) works as a data analyst for Greenpeace, USA.

In addition, two of the authors of this book, Jen Townsend and Steven Klee, are also currently working in data science following careers in academia. Jen Townsend (JT) worked as a tenured community college professor for seven years before transitioning to Microsoft as a data scientist in experimentation (A/B testing). Steven Klee (SK) currently works as an applied scientist for Amazon Web Services. Before that, he worked as a tenured math professor for 9 years.

7.2. Data-Oriented Careers

Please give us a brief description of your work.

SL: I drive research in data-driven approaches to Dynamic Pricing for a portfolio of hotels around the globe. Additionally, I provide business analysis and technical support to a team of Revenue Managers.

JT: I support statistical testing ("A/B testing") of code releases across Microsoft. In practice this means that I (1) research and help develop tools that improve the efficiency of testing/analysis procedures and (2) consult with varying teams who want to take better advantage of statistical methods to measure and improve their product.

BL: I work at a supercomputer facility within a national lab. My job is research-oriented in some ways, but my team's primary mission is to support other people's research on our supercomputers (not primarily our own research agendas). Specifically, I work on applying machine learning/data science to scientific problems. I have especially been working with fluids and dynamical systems applications, so that is pretty math-heavy.

SK: I forecast demand for the Amazon Web Services (AWS) network so that other AWS teams can determine where and when to build new network infrastructure. My job is a mix of data science, model development, and ultimately writing code to deploy models in production.

RS: I currently work for Greenpeace USA as a Digital Performance Analyst.

What path did you take to your current career?

SL: After graduating with a BS in Mathematics (where I completed a thesis in Abstract Algebra), I pursued a PhD in Applied Mathematics and Statistics. A year after receiving my master's degree in this field, I decided to drop out and pursue a career in industry. I started my career with a pure business analyst role in the energy market, but quickly morphed the job into a data science position by tackling problems with data science and machine learning techniques. Business analysis jobs often provide very natural avenues to pursue this. After relocation, I found work supporting the Artificial Intelligence team at a self-driving car start-up, where I acquired many useful data engineering skills. After that team was laid off, I landed at my current role applying data science tools in the hospitality industry.

JT: I definitely didn't aim for a career in data science—the career wasn't even on my radar as something I might be qualified for. After I earned my undergraduate degree in pure mathematics (and having a few CS courses under my belt), I got an entry-level software development position to fill a year before entering my Math/CS PhD program. During my brief stint in industry, I realized I *really* should have taken statistics: it was the math-adjacent field that kept popping up in my work. During my PhD program, I took enough statistics and machine learning coursework to obtain a master's degree. I also realized I loved teaching, and after securing a tenure track position at a community college using my Master of Statistics degree, I left the PhD track. I taught for seven years at the community college and, along the way, I developed some statistics and probability classes that ignited my interest in applied statistics, probability, and machine learning. I got a summer position doing research for the team I now work with full time: this summer opportunity came through a person in my college network. (Networks matter!) At the time of writing this, I've been a data scientist at Microsoft for four years, and recently moved into a management position.

BL: I studied mathematics in undergrad at the University of Notre Dame. I followed a track within the math major that is theory-based and intended to prepare students for math PhDs. In my final year, I thought I would study number theory or combinatorics in a PhD program. I ended up "trying out" applied math via a one-year master's degree at the University of Washington in Seattle. A favorite course was discrete optimization, which built on my interest in combinatorics. I liked applied math and stayed in the department for a PhD. I took more classes in optimization, and my initial research project in optimization exposed me to machine learning. I ended up pivoting to my thesis being more about machine learning (applied to "applied math" sorts of problems, like dynamical systems). I didn't intend on staying in academia, but I surprised myself by staying for two more years as a postdoc, partially due to getting excited about the funded research projects. I then started working in my current position at Argonne. It appeals to me that it's less about "publish or perish" and more like consulting and helping other people. At the time of writing this, I've been at Argonne for two and a half years.

I should mention that I took three computer science classes in undergrad as electives, which ended up being very helpful when I pivoted away from theoretical math. I was also able to participate in math REUs two summers during undergrad, both of which involved programming. Later, I did two internships that further built on my programming skills.

SK: Like Jen, my path to a career as an ML Scientist was not part of the original plan. I have an undergraduate math degree with a minor in computer science, and a PhD in theoretical math specializing in geometric combinatorics. This led to a postdoc and later a faculty position in the math department at a small university.

After about five years in my faculty position, I started to get more interested in data science and machine learning, and I used part of a sabbatical to take Andrew Ng's Coursera course on Machine Learning. I developed and taught a course on baseball sabermetrics and became our department's go-to resource for students who wanted to do senior projects on sports analytics. A few years later, I started to become interested in careers in data science/ML, and I was able to get my foot in the door at AWS through some connections from grad school (+1 to Jen's "networks matter" advice!).

RS: I graduated from Seattle University in 2019 with a Bachelor's in applied math. After graduation in 2019, my professional world was rocked with things like Covid-19 and political unrest and protests. This spurred me into volunteering at Greenpeace in 2020, which ultimately led to a permanent position as a data analyst.

During my time at Seattle U, I worked as a Learning Assistant.[3] Through this experience, I found that I truly enjoyed speaking about math with

---

[3] Learning assistants support students in first- and second-year classes by facilitating study groups and exam prep sessions.

7.2. Data-Oriented Careers

different people. To this day, being a Learning Assistant has been one of my most valuable professional experiences. Explaining concepts like calculus, differential equations, and linear algebra to students not only improved my understanding of those subjects, they also helped me learn how to "speak" math. Learning how to make numbers and math less scary to other students and learning how to create analogies to better explain concepts has given me a big step up in my professional world, where a lot of people are still really scared of numbers! Knowing how to bring data and numerical literacy to others helps me help other people make data-driven campaign strategies that bring better environmental policies.

I also worked on undergraduate research projects, which introduced me to solving real-world problems using math. I focused my senior capstone project on using machine learning to build models of retention rates among minority students. From this research, we found that having diverse faculty and mathematical literature that represents the diversity of the mathematical community, especially in entry-level classes, greatly impacted retention rates. This research also really helped me learn more about things like bias in data and data modeling, which is something I think about every day now in my career!

What coursework was most helpful for your current work?

Our five data scientists were in agreement on this question.

Statistics, probability, linear algebra, and computer science coursework were mentioned by all. Machine learning, data visualization, optimization, graph theory, and differential equations were also mentioned (the first two are not common for undergraduate coursework, but can be taken in master's programs or free online coursework).

What advice would you offer to a mathematics student looking at a career in data science?

SL: Understanding the mathematics behind modern ML solutions impresses people during your interviews. Also, there are skills you don't learn in a mathematics program that can be really foundational: don't underestimate the power of a well-constructed data visualization/dashboard, and learn how to use the command line.

JT: While a math background can be an asset in data science, the burden to demonstrate that is on you. To help interviewers and résumé reviewers see that you've done the work to ensure your math background will translate to effective work in industry:

- Take programming courses. Take statistics. If you can't take these as official courses in your degree, take online training or attend a bootcamp.
- Before any interview, spend some time going through software engineering interview questions in Python and SQL. You may need to train yourself in the fundamentals of these languages, and that's ok! Don't worry if you feel like you only know a little: it's expected that you'll learn more on the job.

- Make sure you communicate your thought process during your interview, as your ability to communicate your logic is often critical in data science.
- Read blogs or watch videos from data scientists working in industry. These resources can help you learn what types of problems data scientists and analysts work on, gain exposure to commonly used techniques, learn the jargon, and understand the realistic constraints and business concerns balanced in their work.

BL: Two things:

- When I was in undergrad, I had an overly narrow view of mathematics and what "I did." For example, during a computer science class I took, there was one lecture on the professor's research. He talked about data mining, which intrigued me, but I didn't think of it as something I could do, because I considered myself to be "a math major," not "a computer scientist." Looking back, it's clear to me that data mining involves both fields. (And actually, my job title is "computer scientist" now!)
- Especially within data science, there is a tendency for the job requirements section in a job posting to include far too many things. If you meet half of them, it's often worth applying!

SK: Statistics. Linear Algebra. Computer Science. Rinse. Lather. Repeat.

Having a good theoretical understanding of what various statistical/ML models do is great, but you also need to demonstrate practical skills, especially if you don't come from a "standard" background for the type of job you're pursuing. It can help to set up a GitHub repo where you show off some polished writeups of projects you have done, even if they seem like toy problems. For example, maybe you have some Jupyter notebooks that use pandas to ingest and clean data, matplotlib to do some exploratory data analysis, scikit-learn to train a model, and all three tools to analyze the results. Also, try solving some Kaggle[4] competitions. You don't need to be the top scorer to have an impressive result (but if you were the top performer, put it on your résumé!).

I'd also second what Bethany said—it's hard for companies to find the mythical unicorn who meets all the criteria of every job posting, and job postings often ask for too much. If you can demonstrate abilities for most of their "need-to-haves" but miss on their "nice-to-haves," it's still probably worth applying. At the same time, if the posting is looking for someone with a PhD and 5+ years experience working with neural networks for natural language processing, and you only know what two of those words mean, then it's probably worth looking elsewhere.

RS: I would recommend getting involved as a tutor or learning assistant if your department has these opportunities. One of the most important parts of being a data analyst is being able to explain what numbers are telling you to people who don't particularly like numbers, which was a big part of my job as an LA. When you have to explain something like the Fundamental

---
[4] https://www.kaggle.com/

Theorem of Calculus to a crowd of students, you find that explaining a graph that supports a strategy you're pitching is much simpler.

A last thing that has really, really helped me in the professional world as a data analyst is having professors who truly ingrained the idea of mathematical justice and ethics in me when I was doing research. Now that I've been doing data analysis work for the last few years, I'm truly thankful to have had mentors who taught me to use data the right way! It's really easy to create biases in your work, or use numbers to back an objective versus just seeing the numbers for what they are. Data ethics is something that not everyone has the chance to learn, and I'm forever grateful that I had the opportunity to know people who wanted to use math to make the world a better place.

7.3 Research in Industry

There are applied mathematics research positions in industry—especially in areas like cryptography and privacy-preserving algorithms, machine learning, computer vision, climate modeling, and more. These positions require a similar background to an academic research position (PhD) but with an additional indication of your ability to work in the constraints of "real" systems, such as using industry data and answering industry problems in your research, demonstrating substantial programming experience, or having done an internship.

When working in industry versus conducting academic research, there are definite tradeoffs. You may be discouraged from pursuing an "intellectually interesting" line of inquiry if it is not relevant for your business needs or has low expected return on your investment. You may not be able to publish certain results (e.g., if the outcome is patentable). More of your time may be spent ensuring your research can be applied and scaled (though this can also be an interesting extra constraint on the research). On the other hand, the pay is often better, you won't need to rely on grants and funding, and you'll get to work on very well-motivated problems inspired by actual issues your company or its customers are facing.

To learn more, we spoke with Chris Swierczewski, who is a machine learning scientist.

Please briefly describe your work.

CS: A machine learning scientist researches new mathematical and statistical techniques to teach a computer how to accomplish complex and sometimes human-like tasks. For example, naming the different objects appearing in an image, determining if a paragraph of text is saying something positive or negative about a subject, or predicting the prices of various grocery store items in five years time. Machine learning scientists also work with software engineers to realize these algorithms in a production setting where thousands to even billions of people around the world may want to take advantage of these new computational tools. In my most recent scientist role at Amazon Web Services (AWS), a cloud computing company, I designed and implemented algorithms for doing topic modeling, nearest neighbors search, and time series anomaly detection which are available for use in various AWS services.

Tell us a bit about the path that led to your current work.

CS: There were several key moments that led me down the path to my current role in machine learning and in computational mathematics, in general. Participating in my high school's FIRST Robotics team was the first time I was able to apply my interest in mathematics and programming to a large team project. Collaborating with area experts and team managers to create a complex object was similar to what my work looks like today. Between this and a summer math camp I became convinced that I wanted to do mathematics. As an undergraduate I took a comprehensive math course load. However, an undergraduate research opportunity on the Sage mathematics software project[5] was what really had a great impact on my career path. Having an incredible senior thesis advisor such as William Stein (the creator of Sage) not only taught me what it was like to do actual math research but his advice and mentorship opened a lot of doors into internships, graduate school, and industry career paths, as well as jump starting a network of research collaborators. The theme of combining mathematics and programming continued in graduate school with my master's thesis in numerical analysis and PhD work in computational approaches to applying algebraic geometry to solving nonlinear partial differential equations. Outside of this coursework I learned about software design patterns and hacked on personal projects. When it came time to apply for jobs I was pleased to learn that many employers were (and still are!) interested in people with one foot in math and another in programming.

Of the classes you took as a student, which ones have been the most useful for your professional career?

CS: I didn't take any machine learning-specific courses during my education. That said, my solid mathematics foundation has helped tremendously in my ability to quickly get up to speed on new problems and jump into a number of different sub-fields of machine learning. For me, there were two series of courses that have paid dividends in the science and engineering halves of my work. A graduate series on linear and functional analysis provided a mathematical bedrock that has allowed me to ingest many machine learning concepts quickly. Another series on numerical analysis made me aware of the challenges that come up in implementing algorithms in a floating point arithmetic environment, thus helping me come up with designs that are fast and numerically stable. Based on my experiences over the past five years I feel like these foundations will continue to be useful, even if I move into non-machine-learning work.

What skills have you needed to develop that were not a main emphasis of your academic training?

CS: There was one technical and one non-technical skill that I needed to develop to be successful in my career. The technical skill I needed to develop on my own were my programming skills. Sure, some applied mathematics or science courses ask you to write MATLAB or even some scientific Python code. However, the level of programming skill learned in these courses

---

[5]SageMath is an open-source computer algebra system, which we also discussed in Chapter 6.

7.3. Research in Industry

alone is insufficient for building our fully-functional algorithm prototypes and especially so for production-level code. Designing robust and useful software was something I had to learn on my own. Of course, I learned much more about writing good code from the engineers I work with in my machine learning job. The non-technical skill that I needed to develop for success was project management. Many masters and PhD students learn some project management while assembling a thesis. However, there is a higher level of organization and detail needed to design scientific projects that address customer needs. You also need to be able to convince various stakeholders, such as engineering teams, product management, and senior leadership, that the project is worthwhile. This all needs to happen while taking a project from a nebulous idea to a fully formed timeline. Unfortunately, I didn't experience many opportunities to practice these skills in school. Therefore, I learned project management on the job by taking on larger and larger projects and by listening to team members who have done this kind of work before.

Are there any jobs in your area for undergraduates right out of college, or do the jobs at your organization tend to require advanced training or experience?

CS: Many machine learning science roles ask for at least a master's degree in a related field such as mathematics, statistics, or computer science. There are just too many tools you need in the tool belt to be an effective research scientist in industry. However, there are many more "data science" as well as "machine learning engineer" roles available to fresh graduates. The terms "research scientist" and "data scientist" can be a little nebulous depending on each company's definitions, but data scientists tend to focus on applying already-available machine learning techniques to very particular problems as opposed to, say, designing general-purpose machine learning tools from the ground up. To use an analogy, think of data scientists as the ones who actually build a house whereas the machine learning scientists are the ones coming up with better power tools. (Though sometimes they're asked to build houses, as well!) Some data science positions are accessible with only an undergraduate degree and experience in this kind of role can lead to a future research scientist role if that's something you're still interested in. Another position available to undergraduates is the "machine learning engineer": a software engineer who works with scientists to design and build out scalable and robust machine learning-related software. Finally, one thing to keep in mind is that many of the larger companies offer undergraduate machine learning internships. These are valuable opportunities to not only experience how research is done in industry but to also get your foot in the door for a future full-time job opportunity.

What advice do you have for math students looking for a research-oriented job in industry?

CS: Make connections with people in the industry you would like to join and exercise them. With your math background, plus some additional programming knowledge, you will already be set up to succeed in a technical interview in a machine learning or similar field. The problem you need to solve, first, is getting those interviews in the first place. Although it doesn't describe the whole picture, the adage "it's not about what you know but who you know" is still true for getting that foot in the door. So while you are acing your coursework, be sure to ask your professors, family, and friends if they or anyone they know works at a company you might be interested in. Reach out to those people for a phone call or lunch meeting and start asking them questions. Hiring managers love it when one of their employees can vouch for a potential new hire instead of combing though a pile of anonymous job applications.

7.4 Careers in Finance

What other career options are there for folks who are good with numbers and data? There is a variety of (often lucrative) roles in finance that students with a mathematics background may be qualified for.

We spoke with Kate Belsky, who has a Master's in Business Administration. Kate is a former math major now working as a Senior Finance Manager for a multi-billion dollar consumer goods company in Los Angeles.

Figure 7.3. Kate Belsky used her mathematics degree to earn an MBA and now works in commercial finance. Photo courtesy of Nathaniel Baimel.

Can you give us a quick summary of the work you do?

KB: I work in commercial finance. In general, our role is to partner with our sales teams to tell the story of what's happening with our volume sales, revenue, trade spend, etc. and help drive decisions with financials. We are responsible for annual planning, forecasting, financial reporting, analysis, and so on.

Tell us a little about the path that led you to your current job.

KB: After completing my MBA, I worked in a few startup and independent contractor roles. That led me to work as an analyst (ultimately a finance analyst) with a sales company. Then, I refocused to a more dedicated Financial Planning and Analysis (FP&A) role.

What classes prepared you for your professional career?

KB: Undergrad courses that taught critical thinking were extremely helpful, math or otherwise. More of the content-specific knowledge around general business practices came through my MBA, which further solidified critical thinking skills.

7.4. Careers in Finance

What skills are important for your work that were not a main emphasis of your education?

KB: A technical proficiency with Excel is key. Many soft skills like emotional intelligence, how to coach and manage people, the ability to build relationships, etc. are also very important.

Are there jobs at your organization for undergraduates right out of college, or do the jobs tend to require advanced training or experience?

KB: There is a rotational program geared for those fresh out of school. There are some other entry-level positions but most require some level of training or experience.

What advice do you have for students of mathematics who are considering a job in finance?

KB: I'll speak to a few things:

- For job-searching: In general, make sure you are tailoring your résumé to the type of job you want. For example, are there key words you can put in your résumé that you frequently see in job descriptions for that type of role? Also brush up on Excel skills. At least in my experience, there is usually some sort of Excel assessment as part of the interview to get an idea of where candidates are at. Additionally, some people are still traditional when it comes to hiring. Write a thank you note (email is fine) after your interview if you can. When you apply and interview for a role, research the company and people who you will be speaking with. Ask them questions about the role and their experience. For example: What do you think are the challenges the company will face in the next few years? What does success look like in this role? What does development of people look like? What are things you would like to see happen for your role/team? You are interviewing the company as much as they are interviewing you, and you want to make sure it's a good fit for both sides.

- For perspective: consider different jobs/roles as opportunities to gain different experiences and learn new things. What are the experiences you want to get? If you ultimately know what you want to do, what are the experiences you need to get there?

- For finance, specifically: there are a number of different types of finance roles that will come with their own set of learning curves and industry-specific nuances. Depending on the type of company or role, the math isn't always the most difficult part, but rather the modeling/assumptions are. As you advance in your career, you'll also need to develop problem-solving, leadership, and communication skills. It is critical to be able to take data and turn it into information, and then communicate that information clearly to different levels of audiences across the organization.

Are you noticing any themes in what you've read so far? Here's one that pops out at us: In addition to mathematics-specific skills and the ability to work with specialized software (like Excel), communication, leadership, and soft skills are very important in many industries! Next, let's take a look at a broader category of professions: those related to research in industry.

7.5 Actuarial Science

Actuaries use data, mathematics and financial theory to quantify and minimize risk. A majority of actuaries are employed by insurance companies, though the actuarial science of risk/uncertainty can be applied outside of insurance as well. This career is accessible with an undergraduate degree in mathematics, though content-specific exams are required to enter and advance in the field.

To learn about preparing for and working as an actuary, we interviewed Micah Lenderman (ML) and Chelsea Lenderman (CL), who worked as Fellows in the Casualty Actuarial Society. Both studied mathematics in college before their careers as actuaries.

How would you describe the core work of an actuary?

ML: The best description I ever heard: "An actuary is like an engineer for an insurance company." Nowadays, "insurance company" seems too traditional. I would expand that to include any business which engages in risk transfer. Actuaries try to shape risk transfer products—by design, by price, etc.—and troubleshoot when things go wrong.

CL: An actuary uses (or manages a team that uses) math and data analysis to analyze risk, solve business problems, and forecast the future. The technical work is often performed in Excel or proprietary software designed specifically for various actuarial methods and techniques. Equally important is being able to clearly communicate one's results to others, both actuaries and non-actuaries.

When you were in college, what do you wish you had known about a career in actuarial science?

ML: I learned about actuarial science my sophomore year of college. In my junior year, I altered my focus from sciences to math and economics. I don't regret that I didn't know about the profession earlier. I enjoyed the breadth in my early education and believe my well-rounded education only helped my actuarial career.

CL: I found out about the actuarial career path and decided I wanted to be an actuary while in high school, so I had a solid head start in my actuarial knowledge. However, having interviewed many college students looking for actuarial internships or entry level positions, I would pass along the following two insights: First, an aptitude for math is important, but if you want to progress beyond a certain level in your career, communication and leadership skills become critical. Make sure you are taking classes and pursuing extracurriculars while in school that will help you develop these softer skills. Second, the field has a good gender balance. For instance, there are many companies within North America that have an equal distribution of genders in actuarial roles, including managerial roles.

7.5. Actuarial Science

Are there programs or societies specifically targeted at actuarial science?

ML: The Casualty Actuarial Society and the Society of Actuaries jointly operate `https://www.beanactuary.org/`.[6] This website offers a lot of information that I wish I had known when I was first learning about the profession. There are a few universities with actuarial-specific programs. The courses at these universities are geared to actuarial science and focused on the early exams. This can be double-edged: some students graduate having passed four exams, but their communication and other skills are lacking.

CL: The two main branches of actuarial science are Life/Health and Property & Casualty. They have separate exam tracks, so you'll need to pick a route to pursue. Learn about the Property & Casualty path from the Casualty Actuarial Society (CAS) website at `https://www.casact.org/`. On the Life/Health side, check out the Society of Actuaries (SOA) website at `https://www.soa.org/`. Depending on your school, there may be an actuarial science major and/or actuarial club.

Which classes are most useful to prepare for an actuarial job?

ML: Upper-level probability and statistics. Less obvious, but useful: courses developing solid communication skills, be it written or oral. With that said, if you are near the end of college, don't eliminate yourself based on your coursework and don't be overconfident either. I have worked with former engineers, a musician, an astronomer, a philosopher, and so on. Ultimately, critical thinking, the ability to communicate and collaborate, and a growth mindset are more important than a B.S. in mathematics.

CL: It's less about classes and more about skills: computational, problem solving, using spreadsheets (namely, Excel), programming, communication, business acumen, teamwork, leadership, etc. You have to like math and be good at it to be an actuary, but an actuarial science major is unnecessary. Get a well-rounded education that includes both technical and non-technical courses. You can learn the specific actuarial methods you need via exams and on-the-job training. Getting involved in extracurricular activities is also a good way to develop softer skills.

Which actuarial exams did you need to get your first job?

ML: At the time I was interviewing, I had two of the nine exams required for Fellowship. A lot of people talk about wanting to be an actuary, but until you have toiled through a couple of exams (or more), then you may not be as committed as you think. Employers know this.

CL: I had passed one exam and sat for the second when I got my first job. I think two exams is the sweet spot to show that you have the technical skills and know what you're getting into exam-wise. Additional time is better spent honing other skills versus adding more exams to your résumé. Many companies offer exam support, like paid study time, covering the cost of exams and study materials, etc. In addition to passing exams, having an actuarial internship experience will go far in helping you land your first job.

---

[6]These two actuarial societies are distinct and serve different sub-industries. You can learn more at `https://www.beanactuary.org/`.

How did you "break in" to the career?

ML: Having two exams got me in the door to interview. Students from our college were not recruited by actuarial companies, but at my first company, a recent alumna and classmate demonstrated that we could compete with anybody. I am sure her success helped me break in, too.

CL: Some alumni from my college that worked as actuaries made an annual campus visit for educational and recruiting purposes. My first actuarial job was with their company.

Do you have any other advice you'd offer to an undergrad student looking at a career in actuarial sciences?

ML: Four things: First, actuarial exams don't care about where you went to college. Harvard? Who cares. Generic state college? Who cares. Your studiousness and motivation matter, not your alma mater. Second, the exam process is long, challenging, and often discouraging. In order to succeed, you must be committed. Third, exams are important, but your work is even more important. While you may get paid study time and increased compensation for passing exams, these are supplemental to your work. Ultimately, you are employed and paid based on your actuarial work product, not exams. Last, but not least, mathematical ability will only get you so far. When you start working with other actuaries, you may be average when it comes to math. Generally, if you want to advance in your career, communication skills and business problem solving skills will differentiate you more than your mathematical aptitude.

CL: Find an actuary who you can job shadow for a day to help determine if the actuarial career path would be a good fit for you. Find out if there are any alumni from your school who are actuaries who would be willing to chat about their experience. Attend an info session given by actuaries visiting your campus or another nearby campus. Take an exam to get a feel for what they're like and to help get your foot in the door. Pursue an actuarial internship or two, perhaps one on the Life/Health side and one on the Property & Casualty side. Finally, the exams provide learning opportunities in your early years as an actuary and act as a barrier to entry, but it's your ability to consistently do your job well—by demonstrating your skills, particularly softer skills, when working on assigned projects—that will set you apart.

7.6 Careers in Software Engineering

Many people who choose to major in math also have an interest in programming and computer science. Of the folks we know with strong interests in both math and CS, several have become software engineers. Their work involves designing, developing, testing, and maintaining software. We spoke with Isaac Ortega, a recent graduate who is now a Senior Software Engineer at Alaska Airlines.

Tell us a bit about the path that led you to your current job.

IO: I was lucky to discover and love computer science early on in high school, which allowed me to continue my engineering education in college.

7.6. Careers in Software Engineering

There, I was able to learn a lot about engineering and mathematics and encountered lots of opportunities to apply my skills. Eventually, I found a role as an intern at Alaska Airlines and have stuck with them ever since.

You were a math major, but you also earned a degree in computer science. Would you say that a major in computer science is essential to getting a software engineering job?

IO: I would definitely disagree with the idea that one needs a formal computer science education to be a successful software engineer. I have had the pleasure of working with many amazing engineers that either majored in non-engineering programs, took six-month boot-camps, or simply skipped college and taught themselves how to be an engineer, and I have found them all to be more than qualified for the role and excellent colleagues. I feel that a formal computer science education helps a lot in understanding the principles and theory behind computer science but, while very helpful in certain situations, those aren't the skills necessary to become a successful engineer.

Of the classes you took as a student, which ones have been the most useful for your professional career?

IO: I think Nonlinear Systems and Modeling and Mobile Software Development were both very helpful classes because they both involved large group projects where we had to collaborate in order to achieve a shared goal while meeting regular milestones and communicating updates to our professor. I feel like those two classes in reflection match the scenarios I find myself in as a professional.

How do you feel your math education prepared you for the job you have now?

IO: I think my math education frequently challenged me to tackle a problem from multiple angles, reassess my assumptions, and fill my knowledge gaps when I tackled a problem or proof that I simply didn't understand. I feel like these are skills that I frequently use in my everyday work, and I am glad that I had so many opportunities to strengthen these muscles as an undergraduate.

You did research as an undergraduate. With the benefit of some years of perspective, how do you think that experience has benefited you?

IO: I think that the research I was able to do as an undergraduate was an incredibly helpful experience. As a part of my research, I got to work with a team of undergraduate students and a professor to discover a topic of our research, define a project that properly communicates our findings, and work on that together over the course of a year, all while communicating our progress and frequently iterating and experimenting with ideas. I was also lucky enough to program an application as part of my research. It was such an exciting application of the problem solving and engineering skills that I had grown in my math and computer science education. I feel that it also required a lot of soft skills such as communication, project/task management, and experimentation/discovery that are difficult to practice in most classes but incredibly applicable to any professional role.

What advice do you have for students who are looking for a job in software engineering such as yours?

IO: As your career progresses, your education will become a smaller and smaller box in your résumé. What companies really want to see is evidence that you can successfully build software. Build a portfolio to show off your skills to as many people as you can! Throw in your big class projects, math research, or personal projects and keep on adding to it! Eventually people will notice, and they will reach out!

Is there anything else you'd like to tell us?

IO: My favorite undergraduate class was Intro to Abstract Algebra I. Group Theory is crazy.

7.7 Government Careers

A major employer of people in the US with degrees in STEM fields is the United States government. We spoke with several folks who, like Bethany Lusch above, decided to go the route of a career in the government. In many government agencies, employees are asked not to share too much information about their work outside of the organization. This is the case for our interviewees here, who wished to remain anonymous. We'll refer to our first interviewee, who works for the Department of Defense, as X. We'll call our second interviewee, who works for the Pacific Northwest National Laboratory (PNNL),[7] interviewee Y. Our third interviewee, referred to as Z, works for an undisclosed Federally Funded Research and Development Center (FFRDC).

Tell us a bit about the path that led you to your current job.

X: I applied for my current job during the fall of my last year of graduate school, but because of how things worked out with the lengthy background check process, I needed to make a decision about an academic job offer before I knew whether or not I would be offered a position with the Department of Defense (DoD). Since I had some flexibility in choosing my start date with the DoD, I worked for a year as an assistant math professor at a private, Catholic liberal arts college.

Y: As an undergrad, I attended some summer programs that introduced me to many different career paths for mathematicians. I learned a lot about government and industry jobs and was intrigued by the government ones most of all. I took an internship at a small government contracting company near Washington DC the summer before my last year in college and really enjoyed it. The problems were challenging, but it wasn't just doing proofs and solving math problems. There was a real application we were trying to tackle. I did a second summer there before grad school. Going into graduate school, I had pretty much decided I wanted to go into government or government contracting. I still pursued pure math, but I was

---

[7]PNNL is one of the many national laboratories run for the Department of Energy. National labs are multidisciplinary science research organizations where researchers do fundamental and applied research for government contracts, primarily DoE but for other departments and agencies as well.

7.7. Government Careers

fairly certain I didn't want to go into academia. I was fortunate to apply for and receive a fellowship from the Department of Homeland Security for three of my five years as a grad student. Through that fellowship, I got the chance to intern at PNNL for three summers. I really enjoyed the people I worked with and the problems I got to work on. When I was getting ready to graduate, I applied for a postdoc position at PNNL and got the job! I transitioned to a staff position about one and a half years into my postdoc, and I've been here ever since. I just celebrated ten years at PNNL!

Z: I entered college thinking I would like to be an aerospace engineer. I had no idea there was math beyond calculus. One night at the beginning of my first year, I went exploring in the math library. I was shocked! Topology? Differential geometry? Category theory? Group theory? What was all this? A helpful fourth-year found me on the floor in the dim light, surrounded by books and a flow chart I was creating to understand how all the math was organized and what I needed to learn first. They laughed, and told me it didn't really work like that, and why don't we start with Naive Set Theory? That began my transition from engineering to mathematics. My transition was complete after my first summer when I had the opportunity to work on a fluid mechanics experiment with a team of engineering majors. Our apparatus had a really oddly shaped tank of water, and we needed to figure out the draining time. Excited to apply the math I had learned that year, I started modeling things out on paper to do the calculation. When I showed my peers, they wisely suggested, "Why don't we just time it?" That is when it clicked. I wanted to be a mathematician. When it came time to think about what I wanted to do after college, I was torn between industry and graduate school.

I was extremely lucky to wind up with both options available, and I decided to take my chance with graduate school. I had been told that it was possible to go from graduate school to industry, but it was much more challenging the other way around. I think this advice was correct. I could never have predicted at the time that a mathematics degree, especially a PhD, would position me to be a part of an exciting early wave of data scientists and machine learning engineers. These were new career options for mathematicians that I didn't start hearing about until I was finishing graduate school.

Enchanted with my thesis work, I used to think working in industry would be a grind. After graduate school, I took a job as a data scientist for a tech company in Austin, TX where one of my fellow graduate students had also taken a job. Far from dull, the work was intellectually stimulating, my peers were amazing, my life was more balanced, and my mental health improved tremendously.

However, I missed math research. After two years, I used my vacation days to attend a Women in Data Science and Mathematics (WiSDM) workshop at ICERM.[8] There, I had the opportunity to collaborate with a group of women on some compressed sensing problems, a field in which I had never

---

[8]Author's note: ICERM is the Institute for Computational and Experimental Research in Mathematics. It is a research institute at Brown University funded through the National Science Foundation.

worked before. We all continued to work remotely after the workshop and published several results. We had the opportunity to work together again the following summer at ICERM and continue the collaboration.

This research work I was engaged in outside of my work as a data scientist paved the way for me to apply for a job at an FFRDC. Now, my career looks exactly like a cross between academic mathematics and industry work! I have an opportunity to use all the mathematics I have ever learned and use both pure and applied knowledge on a regular basis. I also get to learn new math daily.

Of the classes you took as a student, which ones have been the most useful for your professional career?

X: No particular class stands out as being the most useful for my professional career, but I have built on the three programming classes I took as an undergraduate. The critical thinking skills I picked up in both math and computer science classes have also come in handy.

Y: Computer programming has been very useful. I'm not a software engineer, but I do explore data and write algorithms. Having learned the principles and structure of programming as a student has been very valuable. I learned Java and Maple in school and Python on the job, but the major concepts translate between languages. I don't know that I can point to any single math class as being most useful. My overall math degree has been extremely useful though. I learned how to do research, think logically, and structure arguments. All of that is part of my job.

Z: Linear algebra. All the courses I was required to take as an undergraduate wound up being important, and most courses overlap and depend upon each other. However, I do think that linear algebra is so fundamental to computation in mathematics, data science, and machine learning. It is probably the pool of content knowledge I draw from most heavily. Probability and statistics are close behind.

What skills have you developed that were not a main emphasis of your educational training?

X: Programming, software engineering, and leadership are the top three skills I've developed.

Y: Writing and presentation skills. I didn't do much of that in school, but I do a ton of it in my job.

Z: I have become a better software engineer. It was very important that I was exposed to different programming languages, algorithms, and operating systems during my undergraduate coursework, but my ability to engineer software that people care about has been learned on the job. I have also become better at applying mathematics.

Are there any job opportunities for undergraduates right out of college, or do the jobs at your organization tend to require advanced training or experience?

X: Jobs are available at the Department of Defense for all degree levels: undergraduates right out of college as well as people with master's degrees

7.7. Government Careers

and PhDs. There are also summer internships for both undergraduates and graduate students.

Y: There are jobs for undergraduates right out of college if they have the right skills and drive. The people that we hire with a bachelor's degree tend to be really curious, driven, and self-starters. They have done data science projects in classes or on their own, or they've done internships.

Z: In my research area, jobs tend to require an advanced degree and/or extensive experience.

To the extent possible, can you tell us about the types of problems you investigate or have investigated in your government career?

X: Some examples are: figure out how to automate (as much as possible) a manual process, determine how secure an algorithm or a given piece of software is, and detect an anomalous situation.

Y: In the ten years I've been at PNNL, I've worked on many different application areas including computational chemistry, computational biology, power grid, cybersecurity, and knowledge modeling. I use graph theory, hypergraphs, and topological data analysis as my main mathematical tools. For me, the exciting projects are those with a clear application that require some interesting mathematical advance in order to solve. In those cases, I get to make a difference in some real world problem, but I also get to develop some new mathematics.

What advice do you have for students who are looking for a job with a government agency such as yours?

Y: Apply for internships! Don't just apply for one place, apply all over. We get tons of applications and can only hire a fraction of the applicants. Many of the labs are very similar in the kinds of problems they do (at least in the computational sciences), so you'll get a pretty good sense of the type of work that is involved in a government job no matter where you intern.

Z: The interview for my current job was a bit like interviewing for an academic position. Because my brain can only hold so much in working memory at a time, and because my knowledge areas spanned several very disparate contexts (e.g., industry data science, pure algebraic geometry), I planned four mini-talks in addition to my main colloquium talk. I used all my mini-talks when being interviewed by various members of staff! I also tried to learn about which staff members had backgrounds in which areas so that I would know which mini-talk they might like best (I didn't want to start talking about moduli spaces to someone who liked high-performance computing). I also had a few interesting mathematics questions to pose to them (that I had already had a chance to think a little bit about). This was partly so that I could get a sense of how they would be to collaborate with and partly to distract them from asking me too many difficult questions on the spot!

As a job candidate, I remember how demoralizing it could be to get a rejection. As an interviewer, I wish I could have told my candidate-self that there are manifold reasons for rejection that might have zero to do with your interview performance—from the position losing funding, to the nearly random selection that can occur among the top qualified candidates. As difficult as it is, dust yourself off and start prepping for the next interview. There is so much luck involved. Your next interview might just be an amazing fit!

When interviewing, always stay courteous, composed, and professional, and don't give up even if you think you're doing poorly. I've had interviews where I was so flummoxed, I wanted to give up and run out of the building. Then, I get an offer! You just never know. On the other hand, I have seen candidates break down or quit while struggling on things that really wouldn't have been a big deal to us to just talk through, but now we worry about their ability to solve problems. Again, there is so much luck involved, nobody expects you to be perfect, it is not worth putting too much pressure on yourself.

Is there anything else you'd like to tell us about working for your organization or preparing for a career like yours?

X: The DoD isn't looking for expertise in any specific area of mathematics, but more a willingness and ability to work on problems of interest to the DoD. All mathematicians with the DoD need to do a certain amount of programming, but the amount varies based on the specific office. The types of problems on which the DoD works is varied enough that a person can have several careers while working here.

Y: Multidisciplinary science isn't just a tagline for us. Every project I've worked on has included people with a variety of backgrounds: software engineers, mathematicians, and domain scientists. I learn new things all the time, not just new math but applications, too. It's a lot to take in, but it gives us the flexibility to reinvent our careers frequently if we want. We can also be very specialized and work in one area, but we don't have to.

Z: I think investing in pure math up front gave me a broader mathematical arsenal with which to approach applications and the opportunity to make interesting abstract connections. I think it has left more doors open to me in my career than going straight for applied math would have, especially since I did not have a particular applied passion driving me forward. If someone is interested in pure mathematics, they should not be afraid to pursue that passion!

While our three interviewees above work for the US government, other math majors have used their quantitative skills for local government jobs. Marguerite Manela, for instance, works for the New York City Department of Sanitation (DSNY). Here's how Marguerite describes her work.

> "I support a network of local compost processing sites and food scrap drop-off sites throughout New York City. I work to engage NYC residents in active learning about composting and to grow organics recycling initiatives with the goal of providing all New York City residents with opportunities to recycle their organic waste. I am also working to grow DSNY's compost distribution program, which promotes organics recycling in NYC by ensuring that residents and community groups have access to DSNY compost, which is created using New Yorkers' food scraps and yard waste.

7.7. Government Careers

The work I do day-to-day is mostly managerial, relationship building, project management, strategic planning. I also work to set up data intake systems, manage program and financial reporting, and negotiate annual work plans and budgets. I oversee construction projects from a budget and approvals perspective and also support our composting programs' logistical and operational planning."

Figure 7.4. Marguerite Manela lends her quantitative skills to NYC through her work with the sanitation department. Photo courtesy of Marguerite Manela.

Marguerite views her math major as giving her a good foundation for her current job, and she recognizes the benefits of her liberal arts education more broadly.

"Studying math was helpful for not being afraid of numbers and for thinking logically and rationally. I took an environmental analysis class my senior year of college which really helped open my eyes and inspire me to pursue composting and other common-sense solutions to our big issues of waste. Studying a variety of humanities courses has allowed me to excel as a manager by building the ability to take in, process, and build understanding of complex ideas and a variety of perspectives. This was a foundation for becoming a more empathetic person over time, which I believe sets me apart from many."

She also shared with us the technical skills she did not learn as an undergraduate that are or could be useful for her to have for her work.

"My first job out of college was in health benefits consulting, where I worked as an Analyst and learned many skills in Excel and data analysis, but I wish I had learned Access and SQL and, generally, more coding and data analysis skills when I was a college student."

Finally, Marguerite has some sage advice for math majors who are applying for jobs that aren't specifically targeting folks with extensive mathematical knowledge.

"Roll with the 'wow factor' that you'll get if you are a math major—I felt guilty about it early on, but definitely having studied math made a big impact on interviewers who may have had negative stereotypes about math majors, expecting us to *not* have other skills too, like communication, reading, writing, etc. Obviously, I find this stereotype annoying if it persists—math majors are super smart, and we are awesome at many things! But my advice is to build off of that first impression and wow the interviewers with all of the other reasons you're amazing for the job as well."

7.8 Jobs in Education

Jobs in education are as vast as jobs in any other field. We will focus on the K–16 educational system in the United States. In Section 2.3.5, we outlined programs of study that are appropriate for careers in secondary education. Now we will turn our attention to other aspects of educational careers that are worth considering.

The first major question to answer is the level at which you want to teach. Teaching at the elementary level (grades K–5), middle school level (grades 6–8 or 6–9), high school level (end of middle school through grade 12), a two-year college, or four-year college all come with different requirements and challenges.

Broadly speaking, the level of mathematical specialization increases as you progress through the K–16 system. Most K–12 teaching positions require a bachelor's degree, whereas two-year colleges generally require at least a Master's degree and four-year colleges require at least a Master's degree and typically a PhD.

At the K–12 level, a Master's degree or higher can lead not only to deeper subject knowledge and expertise, but also to better pay and a higher position on seniority lists. This equates to better job security: for example, if a school whose teachers are unionized has to make layoffs, they generally start at the bottom of the seniority list.

Some schools offer accelerated Master's programs for future K–12 teachers. These programs tend to have names like Master of Arts in Teaching (MAT) or Master's in Teaching (MIT), and they let you earn a Master's degree and obtain practical teaching experience in one academic year.

At the elementary level, subject specialization is not typically required. At the middle-to-high school level, most states have subject-level certification requirements, which were discussed in Chapter 2. At the high school and college level, most teachers only teach classes in their subject of specialization, whereas at lower levels they will teach a range of classes throughout the day.

For many educators, the population of students they work with is a major consideration. Some teachers prefer the charming innocence of working with kindergarteners or second graders all day, but couldn't imagine having to spend their days dealing with hormonal high schoolers. Others don't want to deal with the pre-pubescent angst of fifth or sixth grade tweens, but they think high schoolers are fascinating because their adult identities are beginning to emerge. Others only want to teach at the college level because they want to focus on teaching and researching in their subject area without having to deal with parents. If you don't know the level where you want to teach, internships or student teaching opportunities can help answer the question for you.

Another consideration is whether you want to teach at a public or private school. Some private schools can be highly specialized or mission-driven, which may be appealing to some future teachers. These can include magnet schools that serve underrepresented groups from a wide geographic area; subject-specific schools that focus on STEM, robotics, or outdoor education; or schools with a religious affiliation that is an important reflection of one's faith. Other teachers may prefer the broad level of education that students receive in public schools. In general, public schools tend to have better pay and better job security/benefits than private schools, which is also a nontrivial factor for many teachers.[9]

---

[9] https://nces.ed.gov/programs/digest/d13/tables/dt13_211.10.asp

7.8. Jobs in Education

To learn more, we spoke with Bridget Klee, who has taught math at the high school level in Washington state since 2008.

Please tell us a bit about your educational background.

BK: I was an undergraduate math major at a private liberal arts school. While in college, I considered studying chemistry and pure math, but eventually was drawn to psychology and education classes. I ended up with a minor in education.

To pursue my teaching certificate, I attended a one-year Master of Arts in Teaching program. The program was twelve months long and included student teaching at two different schools over a period of six months. I feel lucky that I found that program, because I know many teacher certification programs only include eight to ten weeks of student teaching.

Tell us a bit about your current job and the classes you typically teach.

BK: I currently teach at a suburban, public high school with about 2,300 students. For my first few years of teaching in a high school, I taught only Algebra 1 and Geometry. Eventually, I started teaching more upper-level math class like Algebra 2 and Precalculus. I currently teach AP Calculus and Geometry.

I teach five periods a day with fifty-five-minute periods. The courses are year-long math courses.

What classes (math or non-math) or other experiences you had as a student were the most helpful in preparing you for your current career?

BK: As a teacher, the experiences that come to mind first are from my own school experience. My AP Calculus teacher in high school was my most inspirational teacher. I was impressed by the passion and creativity that she brought into every lesson. She also cared about her students both in and outside of the classroom.

While in college, I worked in the Calculus help center and tutored at a local high school. The experience tutoring taught me about working with students who are actively seeking help. My experience student teaching in grad school showed me that there are also students who do not seek assistance and become helpless. I knew this was a group I would work with frequently as a teacher. While student teaching, my mentor teacher helped me involve the parents or guardians in the process to try and engage the student.

What skills have you needed to develop that were not a main emphasis of your academic training?

BK: Some of the skills I developed and honed as a teacher are organization and community building.

For organization, I need to have schedules and lessons prepared for multiple classes and be able to accommodate students who are absent, on field trips, or going on vacation. Time management is something students work on, but as a beginning teacher, I spent hours over the weekend getting ready for the upcoming week. Now I know that I need to set aside time each week to prepare for the next week as a whole and each day to prepare for that teaching day. I used to let students come for extra help anytime that I was in the classroom. Some days I'd even have students waiting at the door when I came in before school. Now I let both students and their parents know when I will be available to give extra help, so I can set aside time to do my own work, too.

During most of my academic training, I was very focused on the math that I would be teaching in my classroom. I did not expect that building relationships with my students would be so important to get them to participate. I now spend time at the beginning of the year working to get to know my students and how they like to learn. We do weekly 'Would you rather quizzes' and share positive weekend updates.

What drew you to teaching high school as opposed to other K–12 levels?

BK: When I was in high school, I worked at a local Children's Museum and an outdoor summer camp. In those experiences, I was able to work with children from elementary to high school age. I found that I built stronger relationships with the older children and enjoyed our connections more. I also knew that I wanted to focus on teaching upper-level math classes and that opportunity was only available in high schools.

What advice do you have for new or aspiring teachers?

BK: My advice to new or aspiring teachers is to have fun and build connections. Bring things that you care about into the classroom, because your students will love to learn more about you.

The more my students know about me, and I know about them, the more willing they are to participate with whatever we do in the classroom.

7.9 Other Career Options

The number and variety of careers out there for people with technical skills is immense. When you think of jobs people can do with a math degree, your mind likely turns to careers in academia/education, industry/data science professions, and government jobs. There are so many more careers, however, that can make use of either the knowledge you'll gain from your degree, the skills you develop, or both. Here, we explore just a few. Of course, this is not an exhaustive list—we've had former math students go on to pursue any number of career options, including beer brewing and human resources management—but this section will give you a sense of the breadth of jobs you might be able to pursue with a degree in the mathematical sciences.

Science writing. Do you like to write? Have you ever taken a creative writing course for fun? Have you considered (or do you have?) a second major in English, communications, or journalism?

Do you love science? Are you curious about how scientists use statistical models to understand the effects of climate change? Do you often wonder about what lies

7.9. Other Career Options

Figure 7.5. Sophia Merow has degrees in mathematics and creative writing and now works as a science writer. Photo: Igor Tolkov. (left) Evelyn Lamb got her start in science writing through an AAAS Mass Media Fellowship. Photo courtesy of Evelyn Lamb. (right)

beyond our galaxy and how we learn about parts of the universe that are so remote from us? Perhaps you have taken a few science courses to try to quench your thirst for knowledge about how things work and how we might innovate to respond to our biggest challenges and opportunities. Maybe you even have a double major in a scientific field.

If the questions above resonate with you, reflecting your interests, one career option you might consider is **science writing**. We know many science writers whose careers began with a math degree. Evelyn Lamb and Sophia Merow—two of our impressive, mathematically-trained science writer friends—shared how studying mathematics led to a career in science writing. Here's what they had to say.

Tell us a bit about your educational background.

EL: I got a BA at Baylor University in an interdisciplinary major that was kind of all over the place. In college, I was especially interested in music theory and math, and it was hard to decide which one to choose for grad school. I ended up going with math, and I got a PhD in math from Rice University in 2012.

SM: I entered Swarthmore College thinking I'd major in math or classics; I graduated with a double major in math and linguistics. Most semesters, I took just one math class, so my math major was pretty light (only one semester of real analysis, for instance). I considered going to grad school in logic or history of math but ended up enrolling in the pure math program at the University of Washington. I left after one quarter, having come to the realization that I really preferred one math class per semester (I'm a liberal arts student at heart!) to three plus a TA assignment. I subsequently earned a master's degree in creative nonfiction writing from Johns Hopkins University.

When did you know you wanted to be a writer?

EL: I didn't know about science writing as a career option when I was in grad school, but the math department secretary would often forward us information about opportunities for grad students, and one of those was the AAAS Mass Media Fellowship, where graduate students in the sciences work in newsrooms, including *Scientific American*, NPR, and the *Chicago Tribune*, for the summer. One fellowship per year is funded by the American Mathematical Society (AMS) and goes to a math grad student. It sounded interesting to me, despite my having no real background in writing or communication, and I applied at the end of my last year in grad school. I was really lucky and got the fellowship that year. I was placed at *Scientific American*, and I loved it. It allowed me to learn a little bit about a lot of topics, which I love. I felt that my work was helping people have positive experiences with math. As most mathematicians know, a lot of people have had traumatic experiences with math in school, and I hoped that I was able to help people possibly enjoy learning a little bit about math, or at least appreciate why mathematicians are passionate about their work.

I had a wonderful mentor at *Scientific American* who really made me see science writing as a career that I could do well and enjoy. I continued in my academic career track to a postdoc at the University of Utah. I continued to do some freelance writing on the side, but after two years, I found that my full-time job was interfering with what I wanted to be able to do creatively with my writing, so I decided to leave the university and do freelance writing-full time.

SM: For a long time I was too risk averse to want to be a writer. I was put off by the lack of a clear path to achieve the become-a-writer goal. But when my post-college get-a-PhD-and-become-a-professor plan fell through, I decided to just go for it and see what happened. I got a copy of "Writer's Market" from the Seattle Public Library and started submitting pieces for publication. Once I had a few clips, I applied to the Johns Hopkins writing master's program. In my professors and fellow students, I saw the many ways to "be a writer," and I began to believe I could be one too.

How did you learn how to write well about technical subjects for general audiences?

EL: I was really lucky in having great editors to work with at *Scientific American* when I was just starting out. They gave me a good foundation. I would sometimes get frustrated with them for not understanding my explanation. A couple of times, I even went over to someone's desk to try to explain something in person. Very kindly, they told me that getting them to understand what I wrote wasn't actually the point. The point is for someone who reads the magazine to understand it. The words on the page are the only way I get to engage with the reader. If a science magazine editor is confused about what I wrote, even if I think it's perfectly clear, it isn't. So I've learned a lot on the job from editors I've worked with, but I've also learned a lot from reading other science writers and social media conversations with people who have read my articles. By learning what my readers found illuminating or confusing, I learn what has worked and what hasn't in my writing.

SM: The Hopkins program had a significant workshop component, so my classmates had to read and comment on my pieces, a number of which were

7.9. Other Career Options

about mathematics. Their feedback helped me (1) wrap my head around what the general public knows about math and (2) begin to appreciate strategies for piquing a general audience's interest about things mathematical. Tutoring math has also proven helpful since it regularly puts me in the position of having to explain mathematical concepts in readily understandable ways.

How did you learn about publishing?

EL: Most of my learning has been on the job, for better or worse. As I mentioned, my mentor at *Scientific American* was wonderful, and I still come to her for advice about my career sometimes. A lot of getting jobs in writing is networking. Because my first position was at *Scientific American*, I started my career with well-connected contacts at a well-known publication. From there, I joined the National Association of Science Writers, which has an annual conference that is great for professional development and networking. I also connect with writers and editors on social media.

SM: I learned a bit about publishing by trial and error and by reading online how-tos about making it as a freelancer. My professors in the Hopkins program—published authors and/or editors all—offered insights, as did other writers with whom I crossed paths (at, for instance, the Creative Writing in Mathematics and Mathematical Sciences workshop at the Banff International Research Station for Mathematical Innovation and Discovery).

Tell us how your math background is key to succeeding in your profession.

EL: Some science writers have no educational background in any science fields, and some like me come to the profession with advanced degrees in a field of science. Both approaches have produced a lot of successful writers, and they have their own pluses and minuses. Having a math background means I have a bit of a shortcut into most mathematics-related topics compared to someone whose last math class was in high school. I know almost nothing about fluid dynamics or homological mirror symmetry, but I do know the kinds of problems mathematicians are interested in and how to ask some of the right questions about a topic to see where their work fits into the big picture. But in some ways, the most helpful thing about having a math background might be the fact that it's made me interested in writing about math! Many science publications are interested in having more math stories, but a lot of science writers are themselves intimidated by math or just aren't interested in finding those stories.

SM: In my experience, mathematicians will happily talk to you about their work if you express an interest in it. If you additionally indicate that you have some familiarity with elliptic curves or non-Euclidean geometry, that you understand on some level what the enterprise of research mathematics is about, they'll be over the moon. My math background (1) enables me to convince mathematician subjects that I'm a genuinely receptive audience and (2) prepares me to intelligently interrogate the information they provide me.

In her interview, Evelyn mentioned that she got her start through the American Association for the Advancement of Science (AAAS) Science & Engineering Mass Media Fellowship. Another AAAS Mass Media Fellow, Leila Slomon, shared her experiences in the program in the *AMS Notices*.[10]

Sophia's "just go for it" attitude is one that's shared by other mathematicians who went on to be science writers. Science writer and editor Susan D'Agostino has the same advice to jump right in. Susan holds a PhD in math and has written for *The Atlantic*, *The Washington Post*, *Scientific American*, *Wired*, and more. She is currently an associate editor at the *Bulletin of Atomic Scientists*. You can find an interview of Susan D'Agostino that appeared in the MAA Math Values blog.[11]

Law. Lawyers need to be adept at analyzing and presenting evidence, forming and communicating sound logical arguments, reading technical writing, and more. Many of the skills that are critical for legal professionals are also essential for doing mathematics. It might not come as a surprise, then, that a degree in mathematics is excellent preparation for a career in law.

Abby Tootell is one example of someone who has parlayed a math degree into a career in law. Abby earned a double major in mathematics and political science from Gettysburg College. From there, she went to the University of Pennsylvania Law School. She finished her law degree in 2020.

> "I think the greatest benefit the math major has had in my legal career is the logical thinking skills I developed from it. My personal statement for law school was about how I viewed reading a judicial opinion in the same way as reading a math proof. Different statutes and cases are like axioms and theorems, facts can be about disputes between people or mathematical statements, and the answer is the holding or a new statement. Being able to think through a series of logical statements to draw the conclusion—or to use a series of statements to convince a reader that your conclusion is correct—is the biggest benefit from my math major.
>
> My math classes were really what helped me prepare for the Law School Admission Test (LSAT) and, I believe, what made my application stand out. I would definitely recommend (a math degree) to aspiring lawyers. I don't think there is a substitute for the logical reasoning skills you develop from a math major, which helped me in studying for the LSAT as well as in all my law school classes, especially writing classes. It also just makes your application stand out and gives you something to talk about in interviews. There's a common joke that lawyers are bad at math, so, at the very least, other lawyers assume I'm smarter than I should get credit for!"

Abby's insights into the connections between math and law are underscored by Molly Stubblefield, who is currently pursuing her JD (meaning *Juris Doctor*, the legal equivalent of a PhD) from the University of Nevada Las Vegas. Molly earned her BS in Mathematics with a Minor in Communication Studies from Western Oregon University. She also completed an MA in Mathematics from University of

---

[10]https://www.ams.org/journals/notices/202001/rnoti-p88.pdf
[11]https://tinyurl.com/h6unmpnz

7.9. Other Career Options

Oklahoma. Reflecting on the relationship between math and law, Molly had this to say:

> "The law is built upon rules and facts—these can be viewed as analogous to theorems, corollaries, and lemmas. When researching a case, similar to writing a proof, you need to find how each of these rules and facts play together. Utilizing these building blocks, you can craft an air-tight argument and help build your case. Having a background in proof writing, problem solving, and critical thinking has helped me immensely in my studies of law so far. Having a challenging case or question in front of me and being able to analytically deduce what would need to be true (and then proving that!) to create a holistic and logical argument is invaluable. I think math majors have the foundational knowledge to become very successful law students and, ultimately, very successful lawyers. The creativity and critical thinking skills that math majors gain through their degree makes them highly qualified to understand the nuanced and challenging landscape that is law."

Figure 7.6. Molly Stubblefield has an undergraduate math degree. She is currently studying law, pursuing her JD. Photo courtesy of MKS Images.

Molly had some good advice about other courses to take to complement a math degree if you're interested in law, too.

> "My minor in Communication Studies helped me build my public speaking, debate, and writing skills. I found it to be a really useful complement to my math degree to give me a well-rounded education. I think the single most useful course I took in all of my formal education so far was my Intro to Improv Acting course. Being able to think on your feet quickly and embracing the spirit of 'yes, and' that is so common in improv is really useful when participating in classroom discussions and debates, interviews, and other networking situations. I recommend it to any student, regardless of their future career goals."

Reform advocacy. Mathematics can help solve some of the world's thorniest problems. Mathematician Moon Duchin, for example, is a national leader in advocating for reform in how we draw congressional district boundaries, reducing gerrymandering in our political system. We interviewed two former math majors, Gabe Schoenbach and Anthony Pizzimenti, who work in Duchin's MGGG Redistricting Lab at Tufts University. Let's hear what they have to say.

Tell us about your educational path.

GS: I went to the University of Chicago and majored in math. I knew before entering college that I liked doing math, but it was only in my freshman year calculus courses that I was exposed to proof-based mathematics, and was immediately hooked. I began to get interested in computer science after taking a discrete mathematics course in my third year, and I ended up taking a bunch of theoretical CS courses and getting a minor in CS by the end of college. I also really liked courses on political/economic theory!

AP: My path to mathematics was pretty roundabout. I started at the University of Iowa as a CS major, but after taking my first proofs-based discrete math course and intro linear algebra, I hopped on the mathematics train and added a math major as quickly as I could. The summer after my sophomore year, I was a fellow at the Voting Rights Data Institute (hosted by the illustrious Moon Duchin and company) and saw how collaborative and fun working on math with others can be—I never looked back! I've always had interests in civics, public policy, and politics as well, so I made sure to solidify my knowledge by taking courses there—and working at the Redistricting Lab, we get a daily dose of all those things. I also took a number of English, Rhetoric, and Gender, Women & Sexuality Studies courses to help round out my understanding of history, culture, arts, etc.

What is your current job title? Explain what sorts of things you do for your work.

Figure 7.7. Gabe Schoenbach works as a research analyst at the MGGG redistricting lab. Photo credit: Miranda Redenbaugh.

GS: I'm currently a Research Analyst at the MGGG Redistricting Lab. Broadly, our lab aims to contribute to the conversation about fair political redistricting: what constitutes a racial/partisan gerrymander, what are good criteria to consider when drawing districting plans, how might different districting priorities conflict with or support each other, etc. Our most important tool is a family of Markov chain Monte Carlo (MCMC) methods that sample from the space of districting plans to help answer these questions. In my work, I tinker with our MCMC algorithms using local search techniques and other methods to sample plans that have desirable qualities (like keeping counties or municipalities intact). I also do a lot of data processing and visualization so that our results can be easily distributed to and understood by districting practitioners.

AP: I'm a Data Scientist at the MGGG Redistricting Lab, where I work on problems at the intersection of mathematics and civics. I'm actively engaged in a number of projects, including math/CS academic research, programming and software development, and loads of data management. Most of the research I work on lies generally in graph theory and algorithms, where I use ideas from combinatorics, algebra, and topology to describe graph-theoretic problems and tell computers how to solve them. I also use many of these same ideas to develop fast, open-source software tools to make typically hard programming tasks—like dealing with large data sets, creating and scoring districting plans, and running Markov chain simulations—easy.

7.9. Other Career Options

What excites you about your work?

GS: My favorite part about this work is how it blends theory and direct application. I'm really interested in the math our sampling algorithms are based on, and our work takes us right up to the frontier of the field—there are tons of fascinating open questions that we get to think about. At the same time, it's easy to see how our day-to-day work can help line-drawers around the country, and it's gratifying to feel like we are making an impact.

AP: I'm particularly motivated by the idea that our work is done in the name of the public good and betterment of our democracy. Solving theoretical problems and using those solutions to improve the lives of others can be rare; being able to contribute to that improvement every day, even a little, is a blessing. I'm also grateful to work with such eminent colleagues, and I wake up in the morning knowing that I'll have learned new things by the time I clock out.

What is the most challenging aspect of your work?

GS: The most challenging aspect of my work is the software engineering side of it. There are so many decisions to be made when writing code to sample or analyze districting plans, and I have to balance a lot of different priorities: speed, correctness, portability/usability by other coders in the lab, etc. While I have formal CS theory education, I developed most of my programming skills on the job, and I feel like I learn better coding practices with each new project.

AP: To be honest, the timelines: redistricting work is really hot right now, especially as the need for redistricting expert work rises with the number of cases in the courts. Turning products around quickly—and when I say quickly, I mean on the scale of hours sometimes—can be extremely challenging. That's why the entire team puts an immense amount of effort into building and documenting software tools that help us get results quickly. That kind of preparation is necessary especially for someone like me who's not the greatest under time pressure.

What qualifications must someone have to do the kind of work you do?

GS: I think there are only two big prerequisites needed for this sort of work. First, a passion for our overarching mission—at its core, fair redistricting is a voting rights issue, and it's much easier to do good work when you're invested in the final goal! Secondly, while it's of course helpful to have some basic math/CS background, I think it's most important to be open to learning new technical skills. The fun thing about programming is that it's basically meant to be learned piecemeal, project-by-project! This job has been my introduction to coding and it's been really fulfilling to gradually pick up new skills as needed.

AP: Being knowledgeable in at least one general-purpose programming language is important, as programming underpins so much of what we do. Furthermore, having a solid foundation in mathematics (particularly discrete math) and computer science fundamentals allows me to meaningfully participate in discussions, especially about complex topics. Beyond academic qualifications, though, a willingness to learn on the fly, share your work with others, be kind, be respectful, do things collaboratively, and remember the humanity of our work are all key to doing well in a data scientist's role.

What advice do you have for current undergraduates who have similar career interests and goals to your own?

GS: I think one important thing current undergraduates could do is to find mentors. They could be professors who you make connections with during office hours, graduate student TAs, or other researchers at your university who have similar interests. While there can be many exciting research and career opportunities out there for you, it can be really hard to know how to find them! The sooner you establish relationships with more senior academics who can get to know you and your interests, the easier it will be for them to help you accomplish your goals.

AP: It may sound a bit cheesy, but make connections with as many people as you can. I made my way into the math(-for-redistricting) space by asking questions, being curious, being vulnerable, and making friends with people I admired. It's important to know that other people want to help you— your professors, your TAs, your departmental staff—and are often there to help you when you need it. As for math-specific advice, find a topic you're really into and just go for it. I've read as many algebraic/topological graph theory books I could find, and I'm so much better off for having done it. Also, building your skill set should happen from the foundation up: always check your understanding of basic principles, and be honest (especially with yourself) when you don't know things – when you get to more advanced topics, fluency with fundamentals is a must.

One reason why Gabe and Anthony went to work for the MGGG Redistricting Lab is that they wanted their work to make a difference in the world. This drive to help others is shared by math majors in our next career category: medicine.

Medicine. Working in medicine requires a strong foundation of scientific knowledge, but it also requires well-honed analytical skills, a capacity to pay close attention to details, the ability to read and understand technical research papers, and a talent for communicating complex information at a level that those without training in medicine can understand. Many of these same skills are ones that math majors develop throughout their course of study. For this reason, medical schools love to see applicants who have a math major in addition to a knowledge base in biology, chemistry, and physics.

To find out more about the path from math to medicine, we started by interviewing Mckenzie Keeling-Garcia as she was finishing medical school and applying to residency programs with the goal of becoming an OB-GYN.

Tell us about your educational journey.

MKG: I took most of my prerequisite courses at a local community college, then transferred to university to finish up my bachelor's in math. I definitely struggled in my math classes and took a while to find my stride, but eventually I did. After this, I completed an SMP, which is a somewhat specialized Master's degree that is designed to help facilitate the transition to medical school from a background other than pre-med. It was probably not strictly necessary to do this program, but it was good experience and allowed me to pursue my goals on my timeline.

How did a background in math help you in pursuing your career in medicine?

MKG: Math explains the way the body works and allows us to take our abstract observations and describe them to others in concrete terms. I use mathematics all day every day without even realizing it, from choosing a treatment plan for a patient, reading about new medical research, or doing my own research in medicine. In Obstetrics and Gynecology, we use mathematical models to determine the probability of successful vaginal birth after cesarean section, to calculate surgical risk, to identify favorable cervices, and more! Another important aspect of having a math background is knowing when to apply a mathematical model and when not to. Some things may not be significant statistically, but are significant to an individual patient, and vice versa.

Figure 7.8. Mckenzie Keeling-Garcia is currently a resident in family medicine. Photo courtesy of Mckenzie Keeling-Garcia.

What advice do you have for current undergraduates who have similar career interests and goals to your own?

MKG: I would recommend a strong math background to anyone who is interested in science and/or medicine. Math allowed me to have some flexibility in the career direction I chose after graduation and gave me more time to find my niche. If a student knows ahead of time that they are interested in medical school, they just need to be sure to add the prerequisite science classes to their schedule as well.

While it took Mckenzie some time to figure out what she wanted to pursue, Spencer Giglio, a first year med student at the time of this interview, always knew he wanted to be a doctor.

Tell us a bit about your background.

SG: I knew from a very young age I wanted to be a doctor. I do not have any family members who are physicians, but some of my father's friends are doctors in the area where I grew up. I found great interest in the stories they told me about medicine and life as a doctor. Over time, I came to realize that medicine was the career path for me after experiences where I shadowed physicians, volunteered in clinics, or worked as a Spanish medical interpreter.

I attended Wake Forest for undergrad where I chose to complete a major in mathematics and a minor in chemistry. I always enjoyed math more than the sciences, so I chose to major in math, rather than biology or chemistry, like many other pre-med students.

How has your background in math helped you in pursuing a career in medicine?

SG: I think that my background in math has helped me in a wide variety of ways, from how I approach problem solving, to my ability to handle complex, multi-step problems. I would say, as a first year medical student currently studying anatomy, that math—and specifically my focus on geometry—helps me visualize the human body in three dimensions. This helps me with radiology, when looking at x-rays and CT scans. I am able to easily relate the two-dimensional structures I see in the scans to the three-dimensional body. I am able to reorient and move parts and pieces around and understand relationships of structures within the body. I believe that my ease with this multidimensional thinking comes from studying topology and geometry and working on reorienting shapes in various ways.

Figure 7.9. Spencer Giglio majored in math instead of following a traditional pre-med path. Photo courtesy of Chase Giglio.

What advice do you have for current undergraduates who have similar career interests and goals to your own?

SG: I would recommend a math degree to students who are interested in pursuing a medical profession. As an applicant to medical school, there are thousands of chemistry and biology majors, and being a math major can really set someone apart from the pack. I also believe that many people choose natural sciences because that theoretically helps prepare someone for medicine, but in my experience, medical school is hard no matter how much background one has in anatomy or physiology. Therefore, being a great thinker and problem solver will be far more beneficial than just having some basic biology background which will likely all be covered within the first day of medical school.

Our next interviewee and former math major went a different route in the medical profession. Martha Cook is a veterinarian. She majored in math with a minor in psychology at Sewanee: The University of the South. As an undergraduate, Martha focused her studies on proof-based mathematics courses, but she also took courses in probability, statistics, and differential equations. It wasn't until after she finished her undergraduate degree that she went back to school to take all of the biology and chemistry prerequisites for veterinary school. She finished her formal education in a dual Doctor of Veterinary Medicine/Master of Public Health program at the University of Tennessee.

7.9. Other Career Options

Figure 7.10. Martha Cook started as a math major and now is a veterinarian. Photo courtesy of R.B. Cook.

How did your math background help you prepare for a job as a veterinarian?

MC: While I don't use real analysis or topology every day in my job as a veterinarian, I do use many of the soft skills that I gained while studying math. The critical reasoning and thought processes used to work your way through a math problem are the same as working your way through a case, from history to exam, diagnostics, differential diagnoses, and treatment. The methodical and step-by-step approach to proofs and other problems is very much applicable to the way of thought required to practice medicine.

I also gained a lot of communication skills while studying math: writing proofs, working through problems with a group, and presenting a problem or proof to my class. Communication is absolutely the most crucial skill for me to have in my current position, so I am very grateful to have gained those skills in college. It is also vitally important to be able to understand journal articles and various studies. I found myself better able to interpret the analysis sections of studies because of my math degree. Someday when I transition to public health and epidemiology, I will be able to do those same interpretations on my own research.

There you have it! We've seen several great endorsements for people interested in math but who want careers in medicine to pursue their mathematical interests in school. Next, let's turn our attention to a completely different kind of career someone might pursue with a mathematics background: a career in art or illustration.

Art and illustration. You've probably heard by now the acronym STEM (Science, Technology, Engineering, and Math), but have you heard of STEAM? The A represents "Art", and is included by some in the acronym as a way to acknowledge the inextricable links between art and other technical subjects. There are a number of academic mathematicians who use their technical expertise to create art, but there are even more artists whose work relies on a sophisticated understanding of mathematical ideas. We spoke to one such mathematical artist, Bathsheba Grossman, about her journey. See Figure 2.1 for a sample of Bathsheba's work.

Coming from a family of acclaimed writers, Bathsheba was the odd one out, expressing an interest in math early on. Her parents didn't know how to support her goal, so they got her a subscription to *Scientific American*. She developed a keen interest in the column written by Martin Gardener who often wrote about puzzles and mathematical games, feeding her fascination with the subject.

Years later, Bathsheba was inspired by her love of math, physics, and computer science to complete an undergraduate degree in math at Yale. Along the way, she took some courses in sculpture to satisfy distribution requirements, and she became hooked.

"I was studying sculpture with this man, Erwin Hauer. He had two currents in his life and, unlike me, he never merged them. He was a figure sculptor and a mathematical sculptor throughout his entire career. All of the classes he taught, though, were from the figure sculpture side, so these are the courses I took from him. It was very enjoyable. It was a lovely break from doing your math homework to go and push some clay around and learn about plaster casting. I think it was after the third course I took with him, there was one day when he took us to his studio. He showed us that he was in fact one of the preeminent mathematical sculptors of the 20th century. It was an epiphany for me. I had no idea that math could be art! And here I was in the presence of someone who was plainly a master at it!"

Figure 7.11. Bathsheba Grossman has captured our imagination with her mathematical art and physical representations of data. Photo courtesy of Bathsheba Grossman.

Inspired, Bathsheba began studying mathematical sculpting with Erwin Hauer. After graduation, she continued to follow this passion, going on to study art in graduate school with acclaimed mathematical sculptor Robert Engman at the University of Pennsylvania. After graduate school, Bathsheba became an early adopter of 3D printing. For a time, she was able to make a living by 3D printing mathematical art, but once 3D printing became widespread, she needed to adapt and apply her creative skills and technical training in other settings. She shifted towards work involving cutting glass with lasers to make breathtaking mathematical and physical objects appear in the glass.

"I wrote the code for laser printing glass from the ground up, which I was only able to do because I have some mathematical aptitude. When the guys with the laser machines described to me what their machines do and the types of lattices that the lasers can handle versus the ones that are too dense, I could work with that because I have some grasp of mathematical methods. Math is a good tool for a lot of different things. Coding is also essential to my work as an artist. It's so helpful to be at the point where you can walk up to a new programming language and pretty much tackle it by the sense of smell."

Bathsheba has turned glass laser etching into a full-time career. She runs a small business called Crystal Proteins which creates 3D illustrations of scientific data inside glass cubes.[12]

"As a data physicalizer, I work with math data, physics data, really any kind of complicated 3-dimensional data that is too fuzzy to be 3D printed and too intractable to be made in any other way. I work with astronomical data. I've drawn a radar scan from caves. I've visualized subcellular x-ray tomography and micro images from organelles in the interiors of cells. I can work with medical images. On the micro level, I've worked with atomic orbitals. On the macro level, I've worked with a large-scale model

---

[12] https://crystalproteins.com

7.9. Other Career Options

illustrating the ways that galaxies aggregate together when you consider millions of galaxies at a time. I can work with all of these things because I have some math background. When a scientist who specializes in one of these fields comes to me and says, 'Hey, I have some kind of data that you've never seen before!' I say, 'Hum a few bars, and I'll fake it!' Because of my background, I can understand what they're saying. I can write a file format converter. I can download specialized scientific apps and understand what they're doing, all because of my math background. A math degree is like a Swiss Army knife. Math is really an ideal major if you don't know what you want to do because there's so many directions you can go with it."

For those who are currently studying math, have some programming skills and an interest in art, Bathsheba has this advice.

"I feel like the place right now where you could have the most fun with mathematics and art without a big startup cost is writing shaders. The more sophisticated the math you do, the more fun you can have. For those who have never heard the term, a 'shader' is a little piece of code that is blindingly fast because it runs on the GPU rather than the CPU of your computer. And this is important because it's running for every pixel of an image that you're going to render. Whenever you see fantastic effects in video games, like turbulence in water, clouds swirling around, or flames, it's because someone has written a shader. There's a large community of people who write these things on a recreational basis."

Bathsheba's path has been along a fascinating, winding road. She has learned how to combine her love of math, science, and programming with her passion for creating art to make beautiful 3D representations of complex objects. Of course, different mathematical artists have a variety of different interests and work with a wide array of mediums. If you're interested in learning more about various types of mathematical art and the artists who create them, the MAA's *Mathematics Magazine* has published a number of artist interviews you can check out: [**19**], [**18**], [**17**], [**16**], [**32**], [**31**], [**30**], [**29**].

What else is out there? For many people, being asked what you want to do for the rest of your life at age 18—or even 22 for that matter—is a perplexing question. How are you supposed to know? You haven't tried enough things and accrued enough experiences to know what really excites you and, importantly, what doesn't! Many students in college explore different majors and minors without knowing where it might take them. You yourself may be a math major who is studying math because you like it, without a serious plan to use what you learn for a STEM-focused career. That's okay.

Meghan O'Neill, who was in just this sort of discovery phase when she was in school, shared the story of her college experience and early career. Here's how Meghan describes her journey.

"I did my undergrad at Vanderbilt University. I double-majored, choosing one major I thought I would like, English, and one I thought I was good at, math. Turns out, I didn't love my English classes, and I struggled in a couple areas of math that I still can't wrap my brain around. But in my heart, I love reading, and I love figuring out math problems, so I feel I made the right decision. At the same time, I minored in Managerial Studies with an emphasis in Corporate Strategy, thinking I might want to go into business. I had a gap year after undergrad and then went to Miami Ad School in San Francisco to do a portfolio program in Art Direction. Now, I work as an Associate Creative Director at a digital advertising agency.

Figure 7.12. Meghan O'Neill majored in math and English and now works as a Creative Director at a digital advertising agency. Photo courtesy of Jenna Horton.

I decided to pursue advertising on a whim, honestly. I had taken an elective for my minor, taught by a former big-wig creative from a top ad agency, and he encouraged me to go to ad school after undergrad. At the time, I was planning on getting my MBA and had a whole path mapped out for myself in fashion and retail, so I didn't consider his suggestion. But after hating my business school interviews, I was left with no plan. I started working retail after graduation while I mapped out a new path, and realized fashion was not the world for me. So, I reached out to my advertising professor, audited a more advanced class of his, and he coached me through my ad school application. I really had no idea what I was getting into, but I took a leap of faith and loved it."

From English to math to business to fashion design to advertising, Meghan took a circuitous route to where she is now. You might think being an Associate Creative Director at an ad agency sounds interesting but wonder what in the world it might have to do with math. We wondered, too. So, we asked Meghan to talk about how her math major has contributed to her success in advertising.

"Firstly, it's a great ice breaker in interviews. When people see my degree in math and my background as a creative, they immediately want to talk about how those two 'types of brains' can possibly exist in the same person. Beyond that, the underlying analytical thinking required for a math degree helps me get to the heart of strategic briefs and understand what audiences truly need out of a product and how to convey that to get them interested in purchasing. I think clients also take you more seriously as a creative when you can speak to and understand their side of things, which is all about the bottom line and crunching numbers."

Whether you've been inspired by an idea for a career in this chapter or not, we hope you've seen in Meghan's story that there's still time to figure it all out. Things may not go as planned during or after college, and sometimes you may not be able to see where you're headed. Don't despair. Keep trying things. Keep exploring. And keep checking in with yourself to figure out which ideas and activities spark your interest and inspire you.

8
Applying for Jobs

If you're leaning towards pursuing a job outside of academia that doesn't require advanced study, you likely have some questions. First, how can you decide which path to follow? Next, if you are settled on pursuing a job in industry, how should you prepare to apply?

In this chapter, we'll begin by discussing some things you'll want to consider in making your decision to apply for certain kinds of jobs. Then, we'll share some advice on the application process.

8.1 What Does Work-Life Balance Look Like for You?

Work-life balance can mean a 40-hour work week to some. To others, it can mean flexibility in where and when you work. Some folks are more concerned with achieving work-life integration: finding synergies and flexibility to support both the work you do and your personal and community well-being. We will use the phrase "work-life balance" to mean the ability to do your job well while still having the energy and time to take care of your other priorities (relationships, family, health, hobbies, travel, sleep, etc.).

There isn't an equation to determine which careers will give you the best work-life balance. A job that energizes you but requires longer hours may give you a better balance than one that drains you despite shorter hours. From our personal experiences working as professors and in industry, we can share a few limited observations, though as we saw in Chapter 7, these certainly are not the only options.

Teaching/academia. At the K–12 level, your work hours will be completely determined by your school's academic calendar with contracted hours for when you are expected to be teaching and attending meetings. At the college/university level, expect to have fairly low "contracted" hours (like 30-35 hours/week) which means that late afternoons and evenings can be used for family and personal priorities.

That being said, many teachers start working again after the work day or week is officially over: they expect to put in 10-20 hours per week of unpaid time in evenings, weekends, and over breaks. At either level, these times outside of work are often the only times available to plan lessons, manage email, complete required trainings, and grade. Others may need to use breaks to work on research (typically at the college/university level) and/or service for department or school.

While there are ample breaks for those in teaching-oriented careers, it can otherwise be hard to take days off during the school year, as it requires advance preparation to arrange for a substitute. The long breaks between terms can be rejuvenating times for rest and vacation for some, but many teachers feel the need to work additional jobs throughout the year—and especially over breaks—in order to get by financially. Others may need to use breaks to complete training, research, and/or service for department or school.

The first years of teaching will often be the least "balanced" as new teachers learn how to navigate their school and their obligations outside the classroom, while preparing lots of new lessons and responding to teaching challenges for the first time. As teachers gain more experience and develop more materials, their effort spent on teaching will go down, but the expectations of them to engage in more service-oriented work will go up.

Industry. What work-life balance looks like in industry is widely variable depending on the sector, company, team culture, and role; each of these can influence expectations around travel, schedule, and velocity. As an oversimplification: compared to stable companies, where a variation on the 9-5 workday is possible in many roles, smaller startup companies and consulting firms have expectations that their employees will work longer hours to respond to the fast pace of innovation or intense client demands.

In the US, industry vacation benefits often start around 2 or 3 weeks per year in addition to official holidays, with that time increasing with your tenure at the company. Some companies have also switched to "discretionary time off" where there is no fixed limit—as long as you are meeting your work requirements and make arrangements with your manager, you can take off time. Generally, there are fewer vacation days than in academia, but there is more flexibility regarding *when* that time off can be taken. In the (albeit limited) experience among the authors of this book, even when it's necessary to work longer hours, it is usually easier to "leave work at work" in industry than it is as a teacher or professor, freeing up mental space and time for other priorities in your life after work hours are over.

It is more common in industry than in teaching for career progressions to have a clear ramping-up of responsibilities and expectations, accompanied by sizable pay increases. Changing roles and companies is far more common in industry, too. Thus, it's common for an industry professional's work-life balance to degrade periodically when they adapt to a new role. It is also not uncommon for industry professionals to seek new positions if the work-life balance in their current role doesn't meet their needs.

8.2 Preparing for the Job Hunt: Finding a Mentor

Since most professors (especially in math) tend to live their lives within academia, their advice and priorities are heavily skewed towards preparing you for the next class or for graduate school in their field. This is not meant to be a disparaging remark against your professors—most professors simply don't have experience outside of academia, and they are most comfortable advising students in areas related to their own experiences. For this reason, it's very helpful to find someone outside of academia who can give you an honest account of their experiences in their industry and suggest what expected skills and qualities you should develop if you want a similar career. Ideally, this mentor will have pursued a career you aspire to, although any professional working in industry will help add perspective.

It can be hard to know where to start looking for a career mentor. Your college career center can often connect you to alumni in various fields. Some professional societies (e.g., "Women in Data Science") have official mentoring programs you can enroll in. You might also consider asking family members or friends to introduce you to people in their networks.

If you can't find a direct personal connection, look for podcasts, videos, or blogs by industry professionals sharing about their career. These are becoming common, and they often outline the required (and overlooked!) skills needed for a given career, discuss expectations for job applications/interviews, and share resources for self-study that can be useful both for students looking to get into the field and for active professionals who want to advance their careers. LinkedIn can be a great place to find and follow professionals sharing content about your desired career.

In addition to searching for jobs on well-known sites like LinkedIn or Indeed, some industries have specialized sites for tracking job postings. For example, USAJOBS[1] is the official employment site for the federal government in the United States. Jobs with the Canadian government can be found on the Government of Canada website.[2]

8.3 When to Apply to Full-Time Industry Positions

The summer before your senior year is generally a good time to ensure you have a draft résumé together. Some larger tech companies (like Microsoft, Google, and Amazon) do mass recruitment/hiring of upcoming graduates between September and January. Smaller companies generally advertise for individual positions less frequently, but they may have more postings in January and July when the fiscal semesters begin. Unlike internships and many academic positions, there are usually not hard deadlines listed for industry job postings. Instead, applications are processed as they arrive until the position is filled. This means it is especially important to apply early if you find a job that interests you. To do this, you need a good draft of your résumé and cover letter, which you can quickly tailor to a specific job. We share advice on how to prepare these below.

8.4 What Skills Are Useful for Non-Academic Careers?

This obviously depends on the career! For most "math-adjacent" industry positions, you'll want to build a solid foundation in coding, statistics, and effective technical communication, with some familiarity with the basics of data analysis, data science, or machine learning. You may learn some of this content during your formal coursework. If not, a background as a math major plus self-study (e.g., through online coursework or a senior capstone project) is also a good path. For some careers, a Master's degree or self study are equally good options for gaining

---

[1] https://www.usajobs.gov/
[2] https://www.canada.ca/

skills. For other careers (often data science) a Master's or PhD is a stated job requirement. Some companies might pay for you to complete graduate coursework or degree programs, so it is also feasible to get an entry-level role right after you earn your undergraduate degree and "level up" to a more advanced role while you work.

8.5 What If I'm Not Qualified for the Job?

As a new mathematics graduate, you will rarely meet all the stated "requirements" of a job posting. Many will ask for 3+ years industry experience, or a swath of programming languages, or a Master's degree in X, or an undergrad in Y (not math). If the job interests you and you think you have the necessary foundation to learn quickly, **apply anyway**. Companies are starting to acknowledge that overly prescriptive job posting requirements can dissuade talented people from applying, but many job postings still have aspirational rather than essential requirements listed.

8.6 How to Write a Personal Statement or Cover Letter

Even though it is geared towards writing a personal statement for an REU, the advice given in Section 3.1 on "How to prepare a personal statement" is applicable here as well: Make your application stand out. Why are *you* qualified and unique? What do you bring to the position? How does the position align with your interests and priorities? For cover letters in industry, there are certain norms in different fields. Folks from your campus career center are often good resources to consult to help ensure that, regardless of field, your cover letter is professional and relevant.

8.7 How to Write a CV or Résumé

A résumé or *curriculum vitae* (CV) both serve the same purpose: to summarize all the work you have done in such a way that it convinces someone you are qualified for the position they have open. Generally speaking, jobs in industry want to see a résumé, while academic positions require a CV. The main difference is length: a person with ten years of experience in academia may have a CV that is 10+ pages long; a person with ten years of experience in the private sector will fit the highlights on a 1-2 page résumé. US government positions tend to be in between, with résumés in the 2-5 page range as they expect more details about previous jobs, as well as additional required information such as citizenship status and/or existing security clearance.

There are many guides to writing a résumé or CV, and we will not try to be the $(n+1)$th comprehensive guide. If you have access to a career center on campus, they will likely have resources for writing and editing of résumés, or you can find guidelines online. In general, you will get advice like:

(1) Keep a résumé short (aim for one page).

(2) If the organization you are applying to has policies or guidelines for résumés: follow them. They may have automation or pre-screening that disqualifies résumés that don't meet their stated requirements in either content or format.

(3) Don't forget to reference your "soft skills," like communication, relationship-building, coordination with teammates, organization, time-management, and self-motivation. These are desirable regardless of the position.

(4) Tailor the résumé to the position. Applying for a teaching position? Your varied tutoring, work with youth, and education-focused coursework is important to highlight. But the same experience might be summarized in one line if you are applying for a position as an actuary, leaving space for you to highlight your statistical coursework. This means you may end up with several variations of your résumé if you are applying for a wide range of jobs.

(5) If you find a position that you really want, tailor your résumé even further to echo the phrasing of the job description. Explicitly include all stated requirements and preferred skills that you have and can strongly demonstrate. If they require "strong analytical and problem-solving skills," you may have a bullet point in your résumé that says something like, "*Developed strong analytical and problem-solving skills* in advanced mathematical studies, such as (fill in relevant courses and a very brief explanation of how they built these skills)." Initial review of résumés is frequently done by recruiters (or algorithms!) who are just looking to validate if you meet the stated requirements—so make it as easy as possible for them to assess this so you can pass to the next round on jobs you are genuinely qualified for.

While the above advice applies pretty broadly, in the next portion we address specific considerations for writing a résumé as a math major.

How do I communicate my math research experience in a résumé? For most non-academic positions, the "what" of your math research doesn't matter nearly as much as the "how." Stating the topic of research and your key result(s) is likely to be entirely meaningless and irrelevant to the reviewer, unless you are applying to a related specialist position. Convey that you were part of a (1) mid-to-long-term project with (2) complexity and uncertainty, where you had to be (3) self-motivated, creative, and persistent, working either (4) independently or collaboratively and which culminated in (5) a presentation to varied audience(s), and (6) a technical report/paper/journal publication of original findings. Reference any coding or technical work learned or executed as part of this research. If it was addressing a particular applied question, briefly describe the application. If it required cross-disciplinary collaboration, mention that. If you worked collaboratively, highlight your particular contributions to results as well as to the coordination, organization, and communication of your group.

If you have publications or have presented at a conference, it's usually worthwhile to list these. Even if the contents don't seem particularly relevant for the position, presentations and publications are indications of your ability to investigate a problem and communicate the findings. Note that the name of the journal or conference may not convey much meaning to your résumé reviewer. Where relevant, you may opt to add "flavor-text." For example: clarify that you presented your research at a *large national mathematics conference (MAA MathFest)* instead of just stating the conference name.

Oh! And did we forget to mention that you should include this content, but only what's relevant for the job, *as succinctly as possible*? Writing a résumé is an exercise in distillation. But take a deep breath, and remember that the résumé is just your attempt to get past initial rounds of screening. It doesn't have to capture your full life experience and skill set. To have fun with using AI interfaces to summarize and review your résumé, see our "Flawed AI résumé review" activity at the end of this chapter.

Do I list the math coursework I've taken? List the coursework that is directly relevant to the job you are applying to. Otherwise you can summarize along the lines of: "Varied advanced coursework in mathematics requiring unguided problem solving, applications of logic, and rapid synthesis of highly technical material."

8.8 Online Presence

It can be important to audit your online presence before you apply for jobs. Many jobs will not or cannot legally review your online presence, but for others, it is a standard part of the hiring process to review your publicly available presence. Ensure that your privacy settings on online accounts are set at the levels needed based on the professionalism you wish to display.

If targeting a private sector position, you may want to create a LinkedIn profile. Many jobs are advertised via LinkedIn, and it provides a professional networking surface. There are a lot of good online articles about how to establish a good, professional LinkedIn profile, and some campus career centers may offer support for setting up an appropriate profile. Whether you like it or roll your eyes at it, LinkedIn is commonly used by companies and recruiters for identifying (and in some cases narrowing down) potential candidates for positions. In the private sector, it is now fairly common for résumés to include a link to the person's LinkedIn profile. If your future job involves programming, you can also set up a portfolio on a site like GitHub to share your code and products, and link that from your LinkedIn profile.

To have a little fun thinking about creating a profile, see our "Social Media Silly Story" at the end of this chapter!

8.9 What If I Apply and Don't Hear Back?

If you send out résumés for an internship or your first full-time position and feel like you're submitting them into a void, you are sadly in good company. Possible reasons include:

- Résumés are sometimes pre-screened by AI or by Human Resources, and so it is quite possible your application will not be reviewed by the team you've applied to. Even if you have skills that translate well to the job, you may not make it through these pre-screenings. This can be especially true if your skills and experiences don't fit the "traditional" pathway into the job.

- Interviews can be a real investment from the company's perspective. Because of this, most try to keep the number of interviews per position low. If a team interviews someone who impresses on a first interview, they may not schedule any other candidates.

- You genuinely don't have the skill set and background the organization is seeking for this role. Perhaps they really need someone who can program in C++

8.9. What If I Apply and Don't Hear Back?

right off the bat, or they are prioritizing candidates who have a portfolio of independently pursued machine learning projects.

- Although you have the capacity to do the work and do it well, you haven't "translated" your experience in math to something that the hiring team can understand. Even if a human looks at your résumé, they will probably take less than a minute to skim the whole thing. Make it easy for them to find the information they are looking for!

- They have prioritized candidates internal to the company or from trusted referrals.

- They could be still reviewing your application, or have hired before your résumé was even received. Companies not sending updates or rejection emails to job applicants is an unfortunately common practice, and if the process feels unfair and inhuman, that's because the process often *is* unfair and inhuman.

Unfortunately, in many cases one of the few ways to ensure your résumé is seriously considered is if you have a direct and solid referral from someone close to the team. A referral often ensures your résumé gets at least a 2-minute review instead of a 30-second skim. You are far more likely to get an interview if you have been vouched for by a current employee of the organization, as this makes you less of an unknown. This is one of the core reasons a strong network beyond mathematicians is useful—a classmate that you studied with in an economics course may land a job at a nonprofit that is looking for a data analyst; a communications major you tutored may end up in marketing at a company that is expanding use of machine learning and needs to hire. You'd be surprised about where important connections might come from! A caveat: in some positions, there are strict fairness practices and HR requirements that prevent or substantially limit referrals. This is more common in public sector positions. But in most circumstances: even if you don't pursue a referral, knowing someone in the position will give you an insight into the skills and needs of the job, and so it is still a substantial advantage.

Getting an interview because you've been "vouched for" can feel like cheating. (And yes, it's a biased and unfair system that can reinforce over-representation of certain demographic groups.) Some folks from demographic groups that don't have as much access to referrals are also less comfortable asking for or using a referral. For example, LinkedIn released a Gender Insights Report covering 2017 and 2018: it reported that women are 26% less likely than male counterparts to ask for a referral, even when they have professional connections at the target company [**24**].

So the truth is: a solid referral often opens the door for you to demonstrate your qualifications. Since breaking into a first job can be extremely hard, we advise you to accept help wherever it is offered, and then make sure you pay it forward (or better yet, advocate to create more equitable practices) once you have your position.

If you don't have the network to offer a referral, don't despair. You can also increase the chances that someone will seriously look at your résumé if you supply it to a company or make direct connection with the company during their recruitment events. These often happen at big name universities but also at industry-specific conferences (especially, but not solely, ones that celebrate and support underrepresented populations). If you can attend the Grace Hopper Conference, for example, you'll have access to a large number of tech industry companies that are looking to hire talented women. The companies' booths often have employees who are there to scout talent and speak with interested attendees. You may be able to chat to learn about opportunities or programs available for new graduates and/or provide a résumé directly at that time (after you spend some time talking with the recruiter: demonstrating your professionalism, communicating your previous experience, and asking thoughtful questions!).

Lastly, if you are not getting any invitations to interview or phone screen, start applying to different types of positions. Your first job may not be with a company or a type of work where you want to end up in your career. That's totally okay! In industry, it's really common to change roles or companies every 3-4 years. Once you've broken into the industry, there *will* be opportunities to change the work you do as well as the team you do it with.

8.10 I Landed an Interview! How Do I Prepare?

First of all: congratulations! That's awesome!

There are likely great resources on interview preparation in your campus's career center or online, but to get started, here are a few tips.

- Before the interview, don't hesitate to ask if they can give you a sense of what the interview will cover. Not only will this help you prepare for the interview, but it also demonstrates your ability to ask questions about a task where you initially have not been given much information.

  Some sample questions include:

  - How many people will you meet on the interview cycle?
  - How long will each interview last, and what will the general format of each one be?
  - Will they be asking technical coding questions? If so, are there preferred programming languages they'd want you to answer in? Will you be expected to write code that will compile, or to write pseudocode on a whiteboard?

- Be prepared to talk about each of the experiences and projects you've listed on your résumé. Prepare a brief summary of each one (1-2 minutes), and be prepared for some standard follow-up questions from the interviewer, like:

  - What challenges did you face? Was it a technical issue or an interpersonal issue?
  - What were your core contributions to the project? (Note: your interview is not a time to be modest. You can be humble while still highlighting what you personally contributed to the group.)
  - What were the big lessons you learned as a result of the project?

- Prepare thoughtful answers to questions about what your greatest strengths and greatest weaknesses are. When thinking about your strengths, be aware that it is common for people to take their best attributes for granted. Ask your friends and mentors what they would say your greatest strengths are to gain some perspective. When you talk about your weaknesses, avoid answers that dodge the question, like, "Sometimes, I care *too much* about my work!"

8.10. I Landed an Interview! How Do I Prepare?

- Look online for examples of "Tell me about a time when..." questions. Try to come up with a few different experiences that can be used to answer these types of questions so that it doesn't seem like you're shining a different light on the same experience in response to three different questions. While it is not universally applicable, the STAR method (meaning: Situation Task Action Result — Google it) is a good tool for organizing your responses to these types of questions so that you efficiently get to the point of what you want to say.

- Practice not just *solving* problems, but *talking through the solution of a problem* (like, "A grocery store chain is evaluating where to locate a new store. How might you build a model to identify top likely locations?") When an interviewer asks you to solve an open-ended design problem like this during an interview, they don't necessarily care about the final solution you give. Instead, they want to observe how you think through the problem. In tech companies, this often is most obvious for coding interview questions, but it applies to all interviews. Here are some specific tips:

    - Restate the question in your own words. This gives you time to think and also gives you a chance to catch invalid assumptions you might be making in your interpretation of the question.
    - Is something in the question ambiguous? Ask about it to get clarity! Sometimes questions are intentionally vague to ensure you can identify and probe into sources of ambiguity.
    - Provide or ask for a small toy example (if possible) so that as you talk through your solution you can explicitly refer to the example.
    - Don't silently think and discard a "dumb" solution as you hunt for something better. Vocalize your thought process: "My first thought is to do X as that will get us a reasonable answer. But it has limitation Y, so now I'm thinking Z."
    - After talking through your solution, pause and reflect: what are the shortfalls of your solution? Are there changes you'd make now that you've had more time to think about the problem?
    - If you genuinely don't know something, just be honest and say that. Don't answer with fancy jargon you don't understand in the hopes that the interviewer won't notice. Interviewers usually have follow-up questions, and trying to fake your way through something you don't understand is generally viewed as a red flag. At the same time, if you know something about the question, share what you do know while honestly assessing the limits of your knowledge. (For example: "I learned in my statistics course that you can find outliers in a one-dimensional data by using the interquartile range, but I don't know how that would generalize if you have to deal with data in five hundred dimensions. That's something I would have to look into.")

- Prepare 2–5 questions that you will ask your interviewer about the position. Use these to check if *you* want to work for the team, and to show your interest. Here are some examples:

    - Ask at least one question that gives the interviewer space to talk about their work or their team's work. For example, "What's a recent project you enjoyed working on with this team?" or "Is there a particular project you have in mind for whomever you hire?"

- Ask about the team's or the organization's culture. If you don't have something specific to ask, a simple "How would you describe the team's culture?" can be revealing. On the other hand, if you want to know something specific, feel free to ask about that. For example, you might ask: "How common is it for people on your team to work on weekends or evenings?", "What support is available for employees to develop new skills?", "What do you see as the qualities and interests that would make someone very happy on the team?", or "Do members of your team tend to work on individual projects or collaborate as a group?"

- Be aware of your rights: interviewing is covered by US employment law and it is illegal for an employer to discriminate against an applicant based on their inclusion in a protected class. Particularly relevant for applicants with a disability: employers cannot legally make pre-offer inquiries about your disability. If you choose to disclose a disability to ensure the interview will accommodate it, an employer must provide reasonable accommodation unless it would be an excessive cost or burden.

8.11 Flawed AI Résumé Review

You may be tempted to use a large language model (LLM) interface for generating a résumé. We aren't going to warn against that. (Who knows what the state of generative AI will be by the time you read this!) Instead, we *would* advise you to use such tools to help you evaluate your résumé once you have it drafted. After all, there's a chance the first gauntlet your résumé must pass will be a stamp of approval from an automated system, so why not see how one interprets your résumé?

Once you have a rough draft of your résumé, supply it to your favorite LLM interface (e.g., ChatGPT, Bard AI, etc.). Then ask some questions. We have included a starter list (both serious and less so) below, and we encourage you to add your own questions to the list.

Remember that—at least as of the date this book's release—the responses from LLMs don't convey truth, just sequences of words that are probabilistically reasonable follow-ups to your input prompt. The purpose of consulting an LLM about your résumé isn't to crush your dreams nor to give you false hope, but rather it is to see a pseudo-objective and flawed response based primarily on your résumé text plus publicly available text databases. Just because the response is highly flawed doesn't mean you can't learn from the response and have fun while you do.

(1) "What university did I attend, and what was my major?" This is a baseline check to ensure the résumé is amenable to whatever automatic text parsers may read it.

(2) "Does my résumé indicate that I'm qualified to be an astronaut?" Unless you have a lot more flight hours than we do, the chatbot *might* be hallucinating if it replies that you are.

(3) "Does my résumé indicate I'm qualified to be an (insert career choice here)?" This response may highlight some skills that aren't coming across strongly in

8.12. A Social Media Silly Story

your résumé or that you genuinely are missing. Check with actual job descriptions to see if these skills are listed as required, and make a plan for how you'll build toward those skills if so.

(4) For a more refined version of the question above, also supply the text of a job posting and ask "Does my résumé indicate that I meet the requirements of this job posting?"

(5) "Suggest a shorter version of (section of the résumé), maintaining the key points."

(6) "Write a haiku based on my résumé."

(7) "Which skills and experiences from this résumé are not relevant for (insert career choice here)?" Note that the response here is helpful because it will highlight skills and experience you may need to re-phrase (or delete) to make the connection to your career choice apparent.

8.12 A Social Media Silly Story

If you'd like to let off some steam as you're prepping to go on the job market, get together with a few friends and come up with words for the parts of speech below. You'll use these words to fill in the blanks in a silly professional, online profile on the next page.

(1) full name of person in room

(2) profession

(3) company

(4) name of a school

(5) academic discipline

(6) adjective

(7) adjective

(8) verb

(9) noun

(10) verb ending in ing

(11) noun

(12) verb ending in ing

(13) verb ending in ing

(14) adjective

(15) verb ending in ing

(16) noun

(17) adjective

A (Silly) Professional Profile

_____ (they/them)
(name of person in room)

Assistant _____ at _____
         (profession)   (company)

About

Graduated from _____ with a double major in mathemat-
                   (school)

ics and _____ . A highly motivated, _____ , and
      (academic discipline)   (adjective)

_____ team player who can _____ well with others.
  (adjective)   (verb)

A _____ -oriented leader with strong _____ skills, as
      (noun)   (verb ending in ing)

evidenced by their accomplishments as President of the _____ Club
                                                            (noun)

and their top performance on the _____ team.
                                (verb ending in ing)

Passionate about _____ and solving _____ prob-
                (verb ending in ing)   (adjective)

lems. Interested in new opportunities to apply their data _____ skills
                                                         (verb ending in ing)

to _____ research and _____ consulting.
      (noun)   (adjective)

9

Graduate School in the Mathematical Sciences

The range of graduate programs in mathematics and mathematically-adjacent fields has as much breadth as the career opportunities that were discussed in Chapter 7. Maybe you want to enroll in a Master's in Data Science or Statistics program to increase your depth of knowledge and become more marketable for industrial jobs in those fields. Maybe you are so enamored with math that you want to have a career as a mathematician or professor, which means you will need a Master's or PhD in Mathematics, Applied Mathematics, Statistics, or something similar. Perhaps you want to teach at the K–12 level but know you will be a better teacher with better pay and more job security after completing a Master of Arts in Teaching program. Or maybe you are similar to some of the folks we heard from in Chapter 7 and want to go into a different field, such as law or medicine, which might require different graduate or professional degrees. The opportunities are endless!

In this chapter, we will discuss graduate programs in mathematics, applied mathematics, data science, or statistics, because those are the areas in which we have the most expertise. So, let's suppose you've decided to apply to graduate school in one of these areas. Now what? You'll need to decide what type of program is right for you, which specific programs you'd like to apply to, how to put together a strong application, and how to prepare for what awaits you in graduate school.

9.1 Graduate Programs: Frequently Asked Questions

What is the difference between a Master's and a PhD program? Master's programs typically involve a significant amount of graduate-level coursework, with a final capstone project. These programs typically last 1–2 years and they may be offered on a part-time basis. Master's programs are also frequently used to redirect a course of study. For example, if you majored in mathematics but decide you want a career in software engineering, you could consider a Master's program in Computer Science. On the other hand, PhD programs often involve coursework and exams in the first year or two, but a major component of the program will involve independent study and doing original research. PhD programs generally take around 4–6 years of full-time work to complete.

Should I apply for a graduate program? If you have a good amount of coursework under your belt from your time as an undergraduate (and not much time has elapsed between your undergrad days and the time you decide to continue your formal education), you may want to directly apply to graduate school. In that case, you'll want to look into either Master's or PhD programs. Or maybe both! Many PhD programs are designed so that you earn a Master's degree along the way, so you do not need to apply for a Master's program separately if your ultimate goal is to earn your PhD. However, if you are interested in a more applied field, like data science, and you plan to work in industry, or if you need a teaching certification, a terminal Master's program might be just right for you.

On the other hand, if it has been a while since you completed your undergraduate degree or if you did not or could not take as many upper-level courses as you need to get into a graduate program, some institutions have dedicated bridge or post-baccalaureate (or postbac) programs to help students prepare for their graduate studies. Postbac programs are programs for people who have earned a Bachelor's degree but would benefit from taking more undergraduate-level courses to prepare for further graduate study or other professional training. Postbac programs are not themselves considered graduate programs, but completing one can help you become a more competitive applicant for graduate programs. These programs often provide help with applying to graduate school, too. A list of such postbac programs can be found on the AMS website.[1]

We recommend looking into postbac and bridge summer programs (which are discussed further in Section 9.6) if it has been a few years since you have thought about undergraduate mathematics or if there are foundational classes, such as real analysis or abstract algebra, that you did not have an opportunity to take as an undergraduate.

How do I get funding for a graduate degree? There are a few common ways to get funding for graduate school other than loans, including

- employer-sponsored tuition benefit programs,
- teaching assistantships (TAships),
- research assistantships (RAships),
- general graduate assistantships, and
- fellowships and scholarships.

Tuition benefit programs are offered by many employers for their full-time employees to continue their education. The range of benefits varies. Employers may have particular academic partnerships, in which case you would attend your employer's approved school (which might even offer classes at your workplace!). Some employers offer flexible tuition reimbursements up to a specified amount—the GI Bill for veterans, military reservists, and active duty military personnel is one example. There are many institutions out there that serve students who have

---
[1] https://www.ams.org/find-graduate-programs

full-time jobs, offering evening or asynchronous classes. This can be a great option if you want some industry experience but are also interested in continuing to learn new skills to advance your career.

Assistantships are funding programs that usually provide a tuition waiver *and* pay graduate students a modest stipend for living expenses in exchange for parttime teaching, research, or other institutional duties. *Note: The applications for these funding programs may be separate from the application to the program itself and can have earlier deadlines.* In mathematics, it is very common for the funding to be allotted in exchange for the grad student teaching, either as the course instructor or as a teaching assistant or grader. Some schools or individual faculty with grant funding offer research assistantships, which fund you to work on a research project that can help you make progress towards your dissertation. For these research assistantships, you often need to have prior research experience. Finally, at the institution level, there may be other funding opportunities offered by professional offices, such as Student Life. For example, dormitories often need graduate students as building resident advisors. These job opportunities are typically offered at the university level, and may not be advertised through a department.

Keep in mind that the type of assistantship you choose may be related to your future goals and past experience. If your goal is to become a professor at a teaching-oriented college, then you'll need to have a fair amount of teaching experience before applying for academic jobs. The best way to demonstrate this experience is to have served as a TA or instructor as part of your graduate studies. If your goal is to become a researcher or a tenured professor at a large research intensive university, you'll want to try to compete for research assistantships—though teaching assistantships can also help.

Note that assistantship funding is often prioritized for PhD students over Master's students, unless the graduate program only accepts Master's students. You should also inquire with any graduate schools you're interested in about how long the funding is guaranteed. Sometimes, funding might be guaranteed for the first year but uncertain beyond that.

Finally, some schools and nonprofits offer fellowships and scholarships you can apply for. Sometimes, they have no additional requirements. Sometimes, they subsidize your teaching load or enhance your pay. An example of a research fellowship is the National Science Foundation (NSF) Graduate Research Fellowship Program (GRFP).[2] This fellowship comes with money for tuition and a living stipend for three years as well as professional development opportunities. Women, underrepresented minorities and people with disabilities are encouraged to apply. We have included an interview with Amzi Jeffs, who was supported by an NSF Graduate Research Fellowship, in Section 9.7.

9.2 Deciding Where to Apply

When you apply for PhD, Master's or postbac programs, it is a good idea to apply to 5–10 programs. But which ones? It depends on what you're interested in and what kinds of support you need to succeed. Do you want to study Number Theory? Biostatistics? Complex Analysis? Data Science? Something else? If you know broadly what field you're interested in, our top recommendation is to

---

[2]https://www.nsfgrfp.org

talk to people who work in that area. Ask them which programs have faculty members actively working in the area you're interested in. Of those, which ones have good track records, retaining a high percentage of their graduate students from the start of the program to graduation and then helping those students find good jobs afterwards?

Perhaps you don't know which field you like best. You might think, for instance, that you like applied math. But maybe you don't know whether you'd like to study tsunamis or the spread of disease. In that case, look for departments that have a broad range of specializations, where you can take a variety of courses to figure out which subject would be the most exciting for you to work in. Ask your professors which programs they know of that would fit the bill.

If the professors at your university don't have the knowledge you are after about graduate programs, there are other ways to find out about programs in your field. Consider networking by attending conferences that are student-friendly. Many national conferences that encourage undergraduate participation (e.g., MAA's MathFest, the Joint Mathematics Meetings, SACNAS Conferences, NAM's Undergraduate MATHFest, Math Alliance's Field of Dreams Conference, Infinite Possibilities Conference) have panels, talks, and events for students to help them learn information about graduate school. These events can also provide you with opportunities to meet faculty from other universities (including ones where you'll want to apply to grad school!). See Section 5.5 for more on your options for attending conferences and networking.

As you are putting together your list of graduate schools to apply to, you'll want to try and learn which programs are going to be supportive of you on your postgraduate educational journey. How can you learn which programs are supportive and which departments have a more competitive or toxic environment? Talk to current graduate students! You can meet graduate students at conferences, but you can also simply email them. Most academic department websites have pages with grad student names and contact information. Consider reaching out to a few current students from each school you're thinking of applying to. Ask them how they like the program. Do they feel supported by their advisors and professors? Do grad students generally get along with each other and work collaboratively? What is the broader departmental community like? What aspects of their program do they feel are the most beneficial? What things don't they like about the program they're in? To get candid feedback, you might even request to have a short phone or Zoom conversation instead of getting information via email. Just make sure in your correspondence with current graduate students that you are respectful of the constraints on their time and express your gratitude for their help in your search. Also, remember to pay it forward by helping other potential grad students find their way when you are a grad student yourself!

In addition to connecting with current graduate students, if there's a faculty member at a university you're applying to who you're particularly interested in working with, consider emailing them. (Have a trusted professor or mentor help you craft this email!) In your message, you'll want to express interest in working with them and find out if they expect to be taking on new graduate student advisees in the next few years. Without talking to them, you may not be aware that the person you want to work with isn't taking research students. Or perhaps they're planning to retire or leave the university soon. So, especially if you want to go to a grad program for the sole purpose of working with a specific faculty member, make sure you connect with that person before going. (See Joe Gallian's *Living Proof* story to find out what can happen when the advisor you want to work with has other plans.)

As you are constructing the list of schools where you plan to apply, it is worth building a range of schools, with some dream schools at the top, solid schools where you think you have a good chance of being admitted in the middle, and some safety schools at the bottom. For all the advice we have presented here, we can still say with confidence that attempting to predict the decisions of a graduate admissions committee is an exercise in futility. A diverse portfolio of schools increases the likelihood that you will be admitted somewhere, even if it isn't your top choice.

In addition, there may be other factors to consider when applying to graduate school. Do you want to attend a school that is located in a large city or in a rural area? Are you willing to apply to schools across the country, or do you want or need to stay in a specific geographic region? Remember that you will work hard during your time in graduate school, but it will not be the entirety of your existence. In fact, the challenges of being a graduate student make it all the more important to be living somewhere with things to do that will make you happy outside of work. If you think you'll be miserable living in a certain part of the country, it probably isn't worth moving there for five or more years just for the sake of being in a graduate program that looks good on paper.

Once you have been admitted, consider visiting campus to see what life is like in the department and off campus. Many departments organize weekends when admitted students can visit and may even have money available to help offset your travel costs. You can often inquire about visiting the department with the head of graduate admissions.

9.3 Applying for Graduate School: Your Personal Statement

One of the most important pieces of your application to graduate school is your personal statement. This is a 1–2 page document that gives the applications committee a sense of who you are beyond what they can read from your transcript and standardized test scores. Much of the advice given in Section 3.1 under "How to prepare a personal statement" still applies here: bring your authentic self to the letter, tell a story, and provide context when necessary. Above all, remember that reading applications is a generally boring and monotonous task for faculty reviewers because most applicants mostly present themselves in the same way. What makes you different? How can you make yourself stand out?

Beyond this, there are some pieces of advice that are more relevant for graduate applications. Why are you applying to *this* specific school? Did you do undergraduate research on multicomplexes arising from Whitney stratifications of matroids, and now you want to work with Dr. Cortez because she is a world expert on the subject? Say that. Are you applying to one and only one school because you already live in town and it is important for you to stay in the area for family reasons? Say that. Are you really interested in the Data Science program at Such and Such University because they are studying social impacts of AI, and you have a second major in technology, ethics, and privacy? Say that. Is there a feature of the program, such as a focus on teacher training, that will help you achieve your long-term career goals? Say that.

It is not always necessary that you have a complete idea of what you want to study in graduate school (as in the case of Whitney stratifications of matroids above), but it is good to have a general idea. For example, you can say "I am interested in your department because I want to study either combinatorics or algebraic geometry, and I know you have strong programs in both fields. I would be interested in potentially working with Professors X, Y, or Z." This gives the committee a sense of your mathematical interests that is more valuable than a vague statement like "I love math and can't wait to learn more about it!"

9.4 Letters of Recommendation

In Chapter 3, we gave advice on letters of recommendation as they relate to REU applications (see Section 3.1, "Letters of recommendation"). The same advice applies here.

9.5 Standardized Tests for Graduate Admissions

In addition to your application to join a specific graduate program, some universities in the US or Canada require that you take the GRE (Graduate Record Examination) in order to be granted admission to the university at large.

There are two categories of GRE that are relevant for math graduate programs: the general exam and the math subject exam. The general GRE is similar to the standardized tests that you might have taken when you were applying for colleges, such as the SAT or ACT. The exam has three sections—Verbal Reasoning, Quantitative Reasoning (high school level mathematics), and Analytical Writing. More information, including sample problems, can be found on the GRE website.[3]

Most universities have a cutoff score on the general GRE that they require for admission, but they do not care as much about your score beyond that cutoff. It is worth spending time working through some practice problems for the GRE so that you will be comfortable with the format.

Some programs also require that you take the GRE Mathematics Subject Test. This is an exam that covers the breadth of the undergraduate mathematics curriculum, with a heavy emphasis on calculus, linear algebra, and differential equations.

The math subject GRE is not an exam to be undertaken lightly. First and foremost, you should determine whether the programs where you are applying *actually* require the math subject GRE. Typically, applied mathematics, statistics, and data science graduate programs are less likely to require the subject GRE than mathematics graduate programs. You will need to check on an institution-by-institution basis to be sure, though.

The subject GRE is offered three times per year—once in September, once in October, and once in April. Practically speaking, if you are planning to apply for admission to a graduate program that will start in the fall, you need to sit for the exam in September or October of the previous year.

Let's get something out of the way right now: the subject GRE is an intense exam. It consists of 66 multiple choice questions to be taken in just under 3 hours. As a back-of-the-envelope calculation, that means you have less than 3 minutes

---

[3]https://www.ets.org/gre

9.5. Standardized Tests for Graduate Admissions

to read and solve each problem. That's not much time, and you need to be well-prepared for it. Some materials to help you prepare can be found on the GRE website, and there are plenty of prep books with practice exams.

Most students, even students who have earned good grades in math classes throughout their college careers, will need to devote a significant amount of time studying for this exam. It is worth making a schedule for yourself, beginning in June or July, where you will set aside a few hours per week to review everything you have learned as a math major. Start with calculus, and make sure you remember all your integration tricks and rules. What are the hypotheses for the Mean Value Theorem? When can you apply L'Hopital's Rule? What is the precise mathematical definition of continuity? How do you find those epsilons and deltas when push comes to shove? Move on to linear algebra and differential equations. Make sure you can identify the techniques that are required for solving a given problem, and then do the algebra. Quickly.

Problems related to upper-level courses can be a bit of a crapshoot. If you haven't taken a class on topology, it isn't worth trying to learn it just for the sake of this exam. There might only be one or two questions related to topology on the exam. Just skip them. No one expects you to get a perfect score, and your time is better spent on the problems that you can solve confidently.

As you are preparing for the GRE, schedule times when you will take practice exams under the conditions of the real test. In particular, this means you are allowed to have pencils and scratch paper. That's it. No music. No headphones. No cell phones. No computers. No books. Reserve a quiet place in the library where you can focus on math, uninterrupted, for 2 hours and 50 minutes. Take the test, grade your work, and honestly assess your performance. Where do you feel confident? What are the areas where you need more practice? It is worth doing this several times in preparation for the exam.

Finally, after all the hard work and energy you devoted to preparing for the GRE, it is easy to feel a bit disheartened when you receive your score. Your overall score will be given as a percentile rank relative to everyone who took the GRE at the same time as you did. This is a pool of students who are applying to graduate programs across the US and Canada, so you are already competing against some of the best math majors in the world. It includes students from the top math programs in the country, students who have already taken graduate-level math classes, and even students who already have a Master's Degree. The pool is full of very talented people. Don't worry if your score isn't in the top 90th percentile. That is a *very* difficult score to achieve. For students who are coming straight out of an undergraduate program with a Bachelor's degree in Mathematics, a score in the 50th percentile is good and a score in the 70th percentile is great.

However, because your score is given as a percentile relative to everyone who took the GRE this year, simple statistics tell us that a third of the people who took the test will score below the 33rd percentile and a quarter will score below the 25th percentile. Maybe this happened to you. This isn't the end of the world, but it also may not help your cause at certain schools. First, it is important to remember that your GRE score is only one component of your overall application packet. Other aspects of your existing body of work, such as internships and undergraduate research experiences, along with your personal statement and letters of recommendation, can certainly outweigh a mediocre or poor GRE score, as is evidenced by Nick Scoville's *Living Proof* story.

At the same time, it could happen that you walk out of the GRE with a keen sense that you have just bombed an exam. One advantage of taking the GRE in September is that you can still edit the list of schools to which you are applying. This isn't to say that you shouldn't still apply to your top schools—it's a hard test and you may still be surprised by your score. However, one strategy that has been successful for some of our students is to look into programs where the Math Subject GRE is not required or is optional. The Math Alliance has compiled a list of such programs,[4] which is an excellent resource. Besides, as Amanda Ruiz points out in her *Living Proof* story, bombing the GRE and not getting into your "dream school" may lead you to a program that ends up being ideal in other ways with plenty of unexpected silver linings.

We should conclude this section by repeating our mantra from Chapter 4: your performance on a very difficult, 3-hour exam that is full of tricks and traps, taken on a Saturday in September is not a measure of your value as a human being. Some would argue that it is hardly a measure of what it takes to be successful as a mathematician. If you crushed the GRE, congratulations. If you didn't, oh well. After the application cycle for graduate schools has completed, the GRE will never matter again in your professional life. It is a hurdle. Nothing more.

9.6 Resources for Graduate Students

If you talk to someone who recently began their studies in a grad program, you're likely to hear them acknowledge that there was a steep learning curve as they moved from undergraduate to graduate-level mathematics courses. Most students struggle with this transition, feeling that they are working tirelessly to learn new, more difficult material while simultaneously filling in gaps in their existing knowledge. There are other factors as well. For example, many students who go to graduate school are used to being top students in their class, but now their graduate class is full of top students, some of whom—intimidatingly—master material much more quickly than they do.

Having gaps in your mathematical knowledge is common. Maybe there were only certain math classes that fit into your college schedule or you spent extra credits getting a second major (which is a strength!), or perhaps the path to your math degree took other twists, turns, or bumps along the way. Some graduate programs may expect you to have taken a core set of courses, for example undergraduate courses in real analysis, complex analysis, linear algebra, and abstract algebra. If you didn't take those classes or if you are not confident in how much of those classes you will remember when you start graduate school in the fall, then you may want to devote part of your summer to reviewing this material.

Fortunately, there are also programs that can help you prepare for grad school by solidifying your understanding of upper-level undergraduate math courses as you prepare for the next phase of your education. For example, the EDGE Summer Program[5] is "a comprehensive program of activities for women entering PhD programs in the mathematical sciences. ... Program participants attend daily lectures in subjects such as Algebra, Measure Theory, Numerical Linear Algebra, and Real Analysis." The Nebraska IMMERSE program, which ran from 2005–15 was a

---

[4]https://mathalliance.org/graduate-record-examination-gre.html
[5]https://www.edgeforwomen.org/

six-week program that aimed to strengthen the preparation of students who were about to enter their first year of graduate school in mathematics. While programs like EDGE are well-established and are likely to continue to run in the future, most programs to support students' transition to grad school, like IMMERSE, typically only run for short stretches of time. Fortunately, at any given time, there are usually a few active programs of their kind. Ask professors in your department and dig around on the internet to see what programs you can find.

Once you're in graduate school, you'll need a community, skills, and knowledge to make it from orientation on Day 1 to walking across the stage in a fancy polyester gown and jaunty hat to collect your diploma n years later. *This* book is not focused on how to navigate the challenges provided by graduate school, but fortunately, there is another book that can help you prepare. We recommend checking out *How to Get Your PhD* by Gavin Brown to get a sense for what lies ahead and learn a variety of strategies to help you succeed.

Brown's book gives advice for all students pursuing advanced degrees, not just in mathematics. For more math-specific resources, the MAA Math Values blog is a good place to start. In particular, check out "Learning Through the Ranks: A Graduate Student Blog," a sub-stream of MAA Math Values.[6]

9.7 Amzi Jeffs and the NSF Graduate Research Fellowship

To learn more about the benefits of the NSF GRFP, we spoke with Amzi Jeffs, who was awarded an NSF Graduate Research Fellowship during his time as a graduate student at the University of Washington.

Figure 9.1. Amzi Jeffs received an NSF Graduate Research Fellowship to help support his studies at the University of Washington. Photo courtesy of Renata Kalnin.

Many PhD programs offer tuition waivers and stipends to their grad students. Given this, why is it useful for students to apply for NSF Graduate Research Fellowships?

AJ: Often, the NSF fellowship stipend is much larger than the stipend offered by a graduate program. For example, when I received my NSF fellowship during my studies at the University of Washington, it constituted a raise of over $6,000 per year! Ideally, all graduate students would be paid a living wage regardless of fellowship status, especially in a city as expensive as Seattle, but until such measures are won universally, an NSF fellowship can make a big financial difference.

It also provides relief from teaching duties. For me, this was generally positive since it allowed me to focus on research, but it also meant that I had to seek out nontraditional avenues for teaching experience, such as teaching for a semester at Cornish College of the Arts,

---

[6] https://www.mathvalues.org/

and volunteering to teach math at the Washington Corrections Center for Women.

Receiving an NSF fellowship is also quite prestigious, and will make you a competitive candidate for any jobs and fellowships that you apply to after graduate school, especially if they are research-oriented.

When should students apply for the fellowship?

AJ: During the first year that they are eligible![7] It may feel like you are not ready to apply, or that the process is intimidating and hard to navigate, but it is very worthwhile to apply as a senior undergraduate student. You will learn a lot of things that will help you in graduate school: how to effectively present your ideas, how to navigate aspects of academic bureaucracy, how to ask for and integrate feedback into your materials, and so on. If you don't receive the fellowship, you can always apply a second time, and it will be much easier with the experience you've gained.

On a smaller scale, it is a good goal to have your materials ready to submit one week prior to the official deadline. This provides some important room for adjustment as you finish your application, and helps avoid any last-minute technical difficulties, which the NSF website is known to experience.

What advice do you have for students who are putting together an application?

AJ: The most important ingredient for a successful application is detailed feedback from a variety of sources (for starters: your advisor, at least one other professor, your friends and classmates, and the tutors at your school's writing center). You should start writing a draft of your application as early as you reasonably can, ideally during mid-September, so that you have time to request and incorporate this feedback.

You should also read as many past NSF GRFP applications as you can. You can find a number of guides and databases with a quick Google search—while working on my application I made heavy use of Alex Lang's website,[8] which includes links to many past applications.

Your application should be as concrete as possible. In your project proposal, you should not only explain what problem you want to solve and why it is important, but what specific steps you will take to begin trying to solve it, and how you can tell whether or not you are making progress. Will you compute examples using code? If so, how will you choose which examples to investigate? What techniques will you need to study for your investigations, and who will you study them with? What special cases of your problem will you try to solve before the general version? And so on.

Likewise, when it comes to the "Broader Impacts" portion of your application, you will need to point towards concrete, realistic plans that you have to improve the mathematical community and society at large (while

---

[7]See https://www.nsfgrfp.org/applicants/applicant-eligibility/ for information on eligibility. The application deadline is typically in late October each year. Students usually apply in the last year of their undergraduate studies or the first year of graduate studies.

[8]https://www.alexhunterlang.com/nsf-fellowship

9.7. Amzi Jeffs and the NSF Graduate Research Fellowship

also recognizing that you alone won't solve every issue). It can be helpful to explain how your concrete activities fit into a larger philosophy of change. For example, in my application I explained that my time teaching at a women's prison was valuable not just because it had a positive impact on my students, but because they had fundamentally different needs in the classroom, and the experience that I and others gained while teaching them made us more effective at discussing math with a broader audience, a vital skill for both teaching and applying our future research results in an effective and just manner.

How have you personally benefited from having an NSF GRF?

AJ: The biggest benefit for me was the increase in salary, which allowed me to escape the uncomfortable and unsafe housing I had during my first years of graduate school. Relief from teaching duties also had a serious impact. I was able to spend a great deal of time on research, and travel more often to give talks and attend conferences. I was also able to participate in activities such as teaching at a women's prison, teaching at an art school, and union organizing in my department, all of which made me a stronger community member and provided valuable experience, but which would have been hard to dig into while simultaneously teaching.

If you don't get it the first time, is it worth applying again?

AJ: Yes, definitely! You will find it easier to write an application the second time you do it, and even if you do not receive the fellowship, the application process provides a chance to develop working relationships with professors in your department and hone your communication skills, both of which will pay dividends during your studies.

Conclusion

Embarking on the journey to become a successful math major requires more than just mastering the theorems and equations found in textbooks. It is a holistic endeavor that requires a combination of curiosity, strategic planning, and perseverance. Your success in mathematics is not solely measured by your ability to solve complex problems, but also by your capacity to communicate your ideas clearly, ask questions when you don't understand something, and appreciate the beauty of the subject.

In these pages, you've learned about building your network within the mathematical community. A successful math major is not an isolated individual working in solitude but a collaborator, who understands the value of teamwork and is able to recognize the unique strengths they and others each bring to the community.

As you move forward in your academic and professional journey, remember that challenges are inherent in the pursuit of mathematical knowledge. There will be moments where you may feel like you're lost without a compass. In those moments, we hope you'll find your people and rely on them. We also hope you'll return to this book to identify resources and find a path forward. Best of luck!

Bibliography

[1] *About us home*, American Mathematical Society, accessed October 24, 2023. https://www.ams.org/about-us/about

[2] *About ASA*, American Statistical Association, accessed October 24, 2023. https://www.amstat.org/about-asa

[3] Duffy Anderson, Matthew Helmer, and Madeline Rue, *Tackling the MCM/ICM*, Math Horizons **31** (2023), no. 2, 26–27.

[4] *About*, Association for Women in Mathematics, accessed October 24, 2023. https://awm-math.org/about/

[5] Heinrich Behrens and Peter Luksch, *Mathematics 1868–2008: A bibliometric analysis*, Scientometrics **86** (2010), no. 1, 179–194.

[6] Jo Boaler, *Mathematical mindsets: Unleashing students' potential through creative mathematics, inspiring messages and innovative teaching*, John Wiley & Sons, 2022.

[7] Tim Chartier, *When life is linear*, Anneli Lax New Mathematical Library, vol. 45, Mathematical Association of America, Washington, DC, 2015. From computer graphics to bracketology, DOI 10.5948/9781614446163. MR3244301

[8] Carrie Diaz Eaton, Hannah C. Highlander, Kam D. Dahlquist, Glen Ledder, Michael Drew LaMar, and Richard C. Schugart, *A "rule-of-five" framework for models and modeling to unify mathematicians and biologists and improve student learning*, PRIMUS **29** (2019), no. 8, 799–829.

[9] Michael Dorff, Allison Henrich, and Lara Pudwell, *Successfully mentoring undergraduates in research: A how to guide for mathematicians*, PRIMUS **27** (2017), no. 3, 320–336.

[10] Michael Dorff, Allison Henrich, and Lara Pudwell, *A mathematician's practical guide to mentoring undergraduate research*, AMS/MAA Textbooks, vol. 63, MAA Press, Providence, RI, 2019. With a foreword by Francis Su; Classroom Resource Materials, DOI 10.1080/10511970.2016.1183248. MR3969934

[11] Della Dumbaugh and Deanna Haunsperger (eds.), *Count me in: Community and belonging in mathematics*, Classroom Resource Materials, vol. 68, American Mathematical Society, Providence, RI; MAA Press, Providence, RI, [2022] ©2022.

[12] Della Dunbaugh, *Calculus abroad: A summer experience in Stockholm*, MAA Focus **43** (2023), no. 5, 6–7.

[13] Carol S. Dweck, *Mindset: The new psychology of success*, Random House, 2006.

[14] Pamela Harris and Aris Winger, *Practices and policies: Advocating for students of color in mathematics*, Self-published, 2021.

[15] Deanna Haunsperger and Robert Thompson (eds.), *101 careers in mathematics*, 4th ed., Classroom Resource Materials, vol. 64, MAA Press, Providence, RI, 2019. For the 3rd ed., see [MR3236893]. MR3970288

[16] Allison Henrich, *Veronika Irvine: The art and mathematics of making bobbin lace*, Math. Mag. **91** (2018), no. 4, 307–309, DOI 10.1080/0025570X.2018.1503465. MR3863788

[17] Allison Henrich, *Robert Bosch: From dominos to traveling salespeople*, Math. Mag. **92** (2019), no. 4, 305–307, DOI 10.1080/0025570X.2019.1624460. MR4009932

[18] Allison Henrich, *Henry Segerman: Visualizing topology*, Math. Mag. **93** (2020), no. 4, 301–305, DOI 10.1080/0025570X.2020.1790969. MR4153213

[19] Allison Henrich, *John Edmark: Art in motion*, Math. Mag. **94** (2021), no. 1, 65–68, DOI 10.1080/0025570X.2021.1843890. MR4213592

[20] Allison K. Henrich, Emille D. Lawrence, Matthew A. Pons, and David G. Taylor (eds.), *Living proof—stories of resilience along the mathematical journey*, American Mathematical Society, Providence, RI; MAA Press, Providence, RI, [2019] ©2019. MR4291889

[21] Brian Hollenbeck and Chad Wiley, *Survey says? Mathematics!*, MAA Focus **43** (2023), no. 5, 44–45.

[22] *About Kappa Mu Epsilon*, Kappa Mu Epsilon, accessed October 24, 2023. https://www.kappamuepsilon.org/about.php

[23] Rachel Levy, Richard Laugesen, and Fadil Santosa, *BIG jobs guide*, Society for Industrial and Applied Mathematics (SIAM), Philadelphia, PA, 2018. Business, industry, and government careers for mathematical scientists, statisticians, and operations researchers; With a foreword by Philippe Tondeur. MR3909430

[24] *Gender insights report*, LinkedIn, accessed October 16, 2023. https://news.linkedin.com/2019/January/linkedin-releases-2019-gender-insights-report

[25] *About MAA*, Mathematical Association of America, accessed October 24, 2023. https://maa.org/about-maa

[26] Sophia Merow, *Problems that feel like play: Sidewalk math as pandemic-era diversion*, Notices Amer. Math. Soc. **68** (2021), no. 5, 798–800.

[27] *Overview*, National Council of Teachers of Mathematics, accessed October 24, 2023. https://www.nctm.org/About/

[28] *What is Pi Mu Epsilon?*, Pi Mu Epsilon, accessed October 24, 2023. https://pme-math.org/what-is-pme

[29] Amy L. Reimann and David A. Reimann, *George Hart: troubadour for geometry*, Math. Mag. **88** (2015), no. 5, 374–376, DOI 10.4169/math.mag.88.5.374. MR3470689

[30] Amy L. Reimann and David A. Reimann, *Anne Burns: mathematical botanist*, Math. Mag. **89** (2016), no. 5, 375–377, DOI 10.4169/math.mag.89.5.375. MR3593662

[31] Amy L. Reimann and David A. Reimann, *Dick Termes: art of the sphere*, Math. Mag. **89** (2016), no. 4, 290–292, DOI 10.4169/math.mag.89.4.290. MR3552770

[32] Amy L. Reimann and David A. Reimann, *Chris K. Palmer: origami in action*, Math. Mag. **90** (2017), no. 5, 380–382, DOI 10.4169/math.mag.90.5.380. MR3738248

[33] *About SIAM*, Society for Industrial and Applied Mathematics, accessed October 24, 2023. https://www.siam.org/about-siam

[34] *About SMB*, Society for Mathematical Biology, accessed October 24, 2023. https://smb.org/About-SMB

[35] *Fostering diversity in science and STEM fields*, Society for the Advancement of Chicanos/Hispanics and Native Americans in Science, accessed October 24, 2023. https://www.sacnas.org/mission-impact

[36] *Who is SOA?*, Society of Actuaries, accessed October 24, 2023. https://www.soa.org/programs/affiliate/soa/

[37] *Spectra*, Spectra, accessed October 24, 2023. http://lgbtmath.org/

[38] Claude M. Steele and Joshua Aronson, *Stereotype threat and the intellectual test performance of African Americans*, Journal of personality and social psychology **69** (1995), no. 5, 797.

[39] Francis Su, *Mathematics for human flourishing*, Yale University Press, New Haven, CT, [2020] ©2020. With reflections by Christopher Jackson, DOI 10.2307/j.ctvt1sgss. MR3971543

[40] *About Us*, The National Association of Mathematicians, accessed October 24, 2023. https://www.nam-math.org/about-us